Renin-Angiotensin Aldosterone System

Edited by Samy I. McFarlane

Published in London, United Kingdom

IntechOpen

Supporting open minds since 2005

Renin-Angiotensin Aldosterone System
http://dx.doi.org/10.5772/intechopen.92503
Edited by Samy I. McFarlane

Contributors
Vinothkumar Ganesan, Jose A. Gomez, Ozlem G. Sahin, Elpis-Athina Vlachopapadopoulou, Myrto Bonataki, Akio Ebihara, A.H.M. Nurun Nabi, Amgad N. Makaryus, Harshith Priyan Chandrakumar, Violeta Capric, Jessica Celenza-Salvatore, Jocelyne Karam, Samara Skwiersky, Sandra Iwuala, Seeta Chillumuntala, Deborah Osafehinti, Volkan Gelen, Abdulsamed Kükürt, Emin Sengul

First published in London, United Kingdom, 2021 by IntechOpen
IntechOpen is the global imprint of INTECHOPEN LIMITED, registered in England and Wales, registration number: 11086078, 5 Princes Gate Court, London, SW7 2QJ, United Kingdom
Printed in Croatia

British Library Cataloguing-in-Publication Data
A catalogue record for this book is available from the British Library

Additional hard and PDF copies can be obtained from orders@intechopen.com

Renin-Angiotensin Aldosterone System
Edited by Samy I. McFarlane
p. cm.
Print ISBN 978-1-83968-286-5
Online ISBN 978-1-83968-287-2
eBook (PDF) ISBN 978-1-83968-458-6

For EU product safety concerns:
IN TECH d.o.o., Prolaz Marije Krucifikse Kozulić 3, 51000 Rijeka, Croatia, info@intechopen.com or visit our website at intechopen.com.

Meet the editor

Dr. McFarlane is a Distinguished Teaching Professor of Medicine/Endocrinology and Associate Dean at SUNY-Downstate, Health Sciences University, Brooklyn, New York, USA. He has extensive experience in clinical and translational research and served as the PI for the largest center in North America in the landmark Diabetes Reduction Assessment with ramipril and rosiglitazone Medication (DREAM) trial. He is the author of more than 400 publications with more than 10,000 citations and an h-index of 46. He also has 270 highly influential citations to his credit, including those in major guidelines by the American Heart Association (AHA), such as stroke guidelines (2018 and 2019), the Scientific Statement from the AHA on Lifestyle and Risk Factor Modification for Reduction of Atrial Fibrillation 2020, and the 2021 SHNE/HRS/EHRA/APHRS Expert Collaborative Statement. He is also the editor of several books on diabetes, hypertension, cardiovascular disease, and related topics. Dr. McFarlane is a three-term member of the National Institutes of Health's National Institute of Diabetes and Digestive and Kidney Diseases (NIH-NIDDK) and served twice as chair of the NIH-NIDDK U01 committee. He also served as chair of the clinical sub-committee of the National Kidney Foundation (NKF) Kidney Early Evaluation Program. Locally, he held several leadership positions including Medical Director of Clinical Research and Program Director and Chief of Endocrinology. He also served as District President of the American College of Physicians. Dr. McFarlane is the recipient of numerous awards and honors including recognition from the United States Army, the US Congress, and the Gold Foundation for Humanism in Medicine.

Contents

Preface

While the Renin-Angiotensin Aldosterone System (RAAS) plays a central role in salt and water homeostasis, it also affects various organ systems including the heart and vasculature, the kidneys, and the nervous system. Evidence indicates that angiotensin II has major deleterious effects on vascular tone, insulin sensitivity, and markers of inflammation and thrombosis. RAAS overactivity is implicated in the pathogenesis of serious and commonly encountered disease entities including hypertension, type 2 diabetes, diabetic nephropathy, left ventricular hypertrophy, congestive heart failure (CHF), and myocardial infarction.

The major pathogenetic mechanisms resulting from RAAS overactivity include activation of the sympathetic nervous system, endothelial dysfunction, and proinflammatory and procoagulant states.

Evidence from basic science and major clinical trials established the beneficial effects of inhibitors of the different components of RAAS such as angiotensin-converting enzyme (ACE) inhibitors, angiotensin receptor blockers (ARBs), and aldosterone antagonists. RASS inhibition is currently utilized in the treatment of hypertension, diabetic nephropathy, and CHF. Inhibitors also demonstrated improvements in outcomes after myocardial infarction and improvement in glucose homeostasis as well as prevention of type 2 diabetes with some agents.

In this book, written by a group of highly experienced scholars, we address the major concepts and topics related to RAAS activation, including the pathogenetic mechanisms underlying the deleterious effects of activated RAAS and the role of local tissue RAAS in various organ systems such as the heart and vasculature, the skeletal muscle, adipose tissues, pancreas, and the angiotensinergic pathways in the brain. Cutting-edge information addresses the needs of a wide range of readers including medical students, clinical practitioners, and basic science investigators alike. This book bridges the gap between basic science and clinical practice regarding the RAAS system, which is imminently critical and highly relevant to today's practice of medicine. Finally, with data emerging from the COVID-19 pandemic indicating overrepresentation of people with diseases associated with RAAS activation such as hypertension, chronic kidney disease, and diabetes, the role of RAAS activation and RAAS inhibition in the pathogenesis and clinical outcomes in COVID-19 has garnered a great deal of interest. In this book, we dedicate a chapter to this topical and highly critical subject.

Samy I. McFarlane, MD, MPH, MBA, FACP
Distinguished Teaching Professor and Associate Dean,
Internal Medicine Residency Program Director,
College of Medicine,
Department of Medicine,
Division of Endocrinology,
State University of New York,
Downstate Health Science University,
New York, USA

Section 1

The Renin-Angiotensin Aldosterone System: Pathophysiologic Insights

Chapter 1

The Role of the Renin-Angiotensin-Aldosterone System in Cardiovascular Disease: Pathogenetic Insights and Clinical Implications

Violeta Capric, Harshith Priyan Chandrakumar, Jessica Celenza-Salvatore and Amgad N. Makaryus

Abstract

Increased attention has been placed on the activation of the renin-angiotensin-aldosterone system (RAAS) and pathogenetic mechanisms in cardiovascular disease. Multiple studies have presented data to suggest that cardiac and arterial stiffness leading to adverse remodeling of both the heart and vasculature leads to the various pathological changes seen in coronary artery disease, heart failure (with preserved and reduced ejection fractions), hypertension and renal disease. Over-activation of the RAAS is felt to contribute to these structural and endocrinological changes through its control of the Na+/K+ balance, fluid volume, and hemodynamic stability. Subsequently, along these lines, multiple large investigations have shown that RAAS blockade contributes to prevention of both cardiovascular and renal disease. We aim to highlight the known role of the activated RAAS and provide an updated description of the mechanisms by which activation of RAAS promotes and leads to the pathogenesis of cardiovascular disease.

Keywords: cardiovascular disease, coronary artery disease, heart failure, hypertension

1. Introduction

Cardiovascular disease is the leading cause of death in men and women in the United States and throughout the world [1]. Current efforts are focused on decreasing the burden of death due to atherosclerosis and cardiac disease overall. Increased attention has been placed on the activation of the renin-angiotensin-aldosterone system (RAAS) and pathogenetic mechanisms in cardiovascular disease. The RAAS system effects blood pressure control and electrolyte and fluid balance and therefore plays a significant role in cardiovascular hemodynamics [2–4].

Classically, it is known that angiotensinogen is cleaved by renin to form angiotensin-I (Ang I), which is then converted to angiotensin-II (Ang II) by angiotensin converting enzyme (ACE), however other peptides and products of this axis have

IntechOpen

been shown to play a role in the development of cardiovascular disease [3, 4]. It is thought that two of these products (angiotensin 1-7 and angiotensin 1-9) may have counterregulatory effects on the development of atherosclerosis and cardiovascular disease [4]. Although the role of angiotensin II is understood more clearly, these peptides provide other targets by which the RAAS system can be utilized to prevent atherosclerosis.

Overactivation or pathologic activation of the RAAS system, specifically angiotensin II, has been shown to play a specific role in endothelial dysfunction, inflammation, intense vasoconstriction, increased vascular and cardiac hypertrophy, fibrosis and the development of atherosclerosis [2–5]. Multiple large investigations have shown that direct inhibition of the effects of angiotensin II via angiotensin converting enzyme inhibitors (ACE-I) and angiotensin-receptor blockers (ARB) improve mortality, prevent renal disease and decrease cardiovascular events in this subset of patients. Additionally, some studies have shown that utilization of both ARB and ACE-I may have cumulative effects on inhibiting the adverse effects of an overactivated RAAS system [6, 7].

We aim to highlight the known role of the activated RAAS and provide an updated description of the mechanisms by which overactivation of RAAS promotes disease and provide a summary of the clinical implications of RAAS inhibition in cardiovascular disease.

2. Overview of the RAAS system

The RAAS system has several moving parts, with different organ systems stimulating its activation and suppression. Renin, the active form of prorenin, is secreted by the granular cells of the kidney. Although renin's role is that of an enzyme, its means of expression are more hormonal. Renin's production is stimulated by hypotension, hyponatremia, and decreased sympathetic activity. Renin is responsible for cleaving angiotensinogen, a protein produced in the liver. Angiotensinogen is regulated via thyroid hormone, steroids, and levels of circulating angiotensin II. Angiotensinogen is cleaved into angiotensin I, which is further converted into angiotensin II by angiotensin converting enzyme [3, 4].

RAAS key players are composed of renin, angiotensin I & II, and angiotensin converting enzyme located in the heart atria, conduction system, valves, ventricles, coronary vessels, fibroblasts and myocytes [8, 9]. Ang II is the effector hormone playing a pivotal role in the cardiac RAAS and has a widespread effect throughout the body, targeting different mechanisms of action.

Ang II acts via the angiotensin receptors mediating the following actions [9, 10]:

1. Cardiovascular system - vasoconstriction, increased blood pressure, increased cardiac contractility, vascular and cardiac hypertrophy

2. Renal system - tubular sodium reabsorption, inhibition of renin release

3. Sympathetic nervous system stimulation

4. Aldosterone synthesis through adrenal cortex

5. Cell growth and proliferation, inflammatory response, and oxidative stress.

Angiotensin converting enzyme 2 (ACE 2) is involved in the degradation of Ang II to Ang (1-7) and Ang (1-9), which provide a relative vasodilatory effect

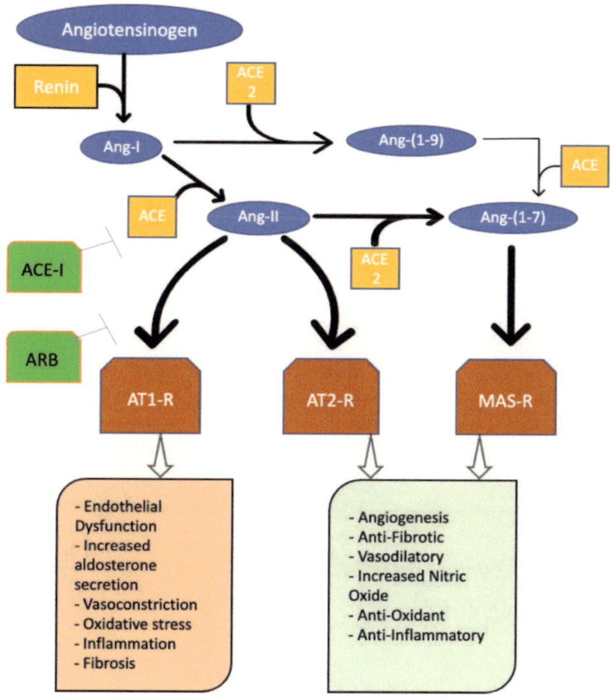

Figure 1.
Schematic of the RAAS as it pertains to angiotensin II and angiotensin (1-7) (Ang-(1-7)) and their counter-regulatory effects via angiotensin receptors 1 and 2 (AT1-R and AT2-R respectively) and MAS receptor (MAS) [5, 6, 11]. Abbreviations: ACE-I (angiotensin converting enzyme inhibitor), ARB (Angiotensin-II receptor blocker), Ang (1-9) (angiotensin 1-9), ACE (angiotensin converting enzyme).

as outlined in **Figure 1**. ACE 2 is restricted to vascular endothelial cells, arterial smooth muscle cells, myofibroblasts, carotid arteries and renal tubular epithelium [8–10]. The effects of Ang II, Ang (1-7) and Ang (1-9) have been uncovered in the past several years, specifically their role in hypertension, endothelial damage, and cardiovascular disease [5, 6, 9, 12]. The role of Ang (1-7) and Ang (1-9) is further outlined in **Figure 1** as they pertain to the pathophysiologic changes in the cardiovascular system.

3. Pathogenic insights

3.1 Atherosclerosis and endothelial dysfunction

Endothelial dysfunction is thought to be a precursor to atherosclerosis, or the thickening and stiffness of vessels. This damage often cultivates in an atherosclerotic plaque, which is a fibrin and cholesterol contained structure that deposits on the inner lumen of blood vessels and can impede oxygen delivery to tissues and organs. Endothelial damage and inflammation allow for the migration of monocytes and macrophages to the site of injury and the formation of foam cells [13–15]. Additionally, stimulation of inflammatory mediators also promotes smooth muscle cell (SMC) thickening, stiffness of vessels and forms a fibrous cap on the atherosclerotic plaque (**Figure 2**) [16]. The pathophysiology of plaque development is very closely tied to RAAS as Ang II plays a key role in these pathophysiologic changes.

Ang II acts on the AT1 and AT2 receptors (AT1-R and AT2-R) causing arteriolar vasoconstriction, and inflammation through generation of reactive oxygen species

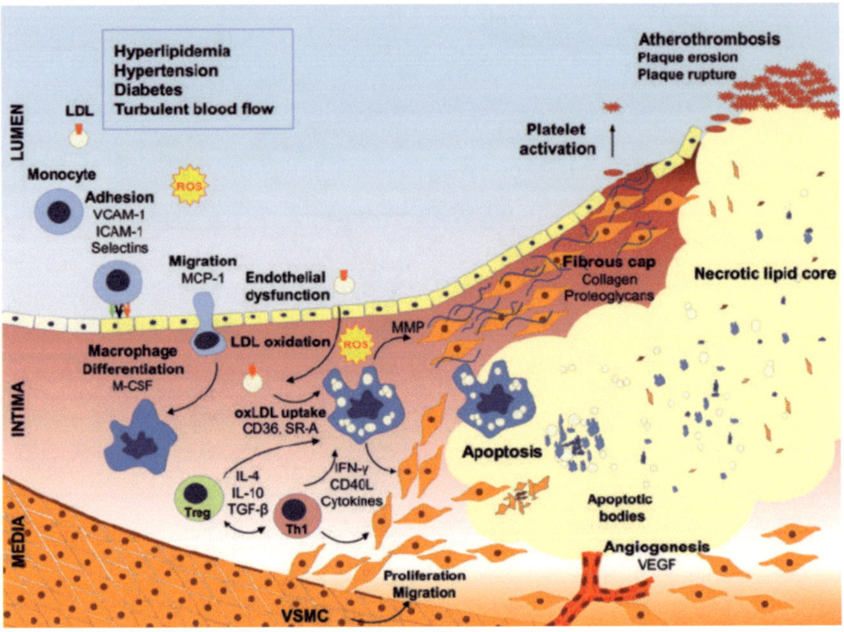

Figure 2.
A schematic depicting the dynamic changes involved in the formation of an atherosclerotic plaque [16]. Abbreviations: ROS, reactive oxygen species; ICAM-1, intracellular adhesion molecule 1; IFN-c, interferon-gamma; IL, interleukin; LDL, low-density lipoprotein; M-CSF, macrophage colony-stimulating factor; MCP-1, monocyte chemoattractant protein 1; MMP, matrix metalloproteinase; oxLDL, oxidized LDL; SR-A, scavenger receptor A; TGF-b, transforming growth factor beta; VCAM-1, vascular adhesion molecule 1; VEGF, vascular endothelial growth factor; VSMC, vascular smooth muscle cells. Reproduced with permission from Mary Ann Liebert, Inc.

(ROS), proinflammatory transcription factors such as nuclear factor kB (nf-kB), and the proliferation of smooth muscle cells contributing to atherogenesis [17, 18]. Activated nf-kB increases inflammatory mediators including interleukin-6 (IL-6), monocyte chemoattractant protein-1 (MCP-1) and platelet derived growth factor (PDGF), all of which mediate inflammation, endothelial damage and monocyte migration and adhesion leading to fibrosis [6, 18].

Ang II induces NF-kappaB (NF-kB) and inflammation through its binding to AT1-R. This has been demonstrated extensively as AT1-R blockers have shown to significantly decrease inflammation. Induction of NF-kB leads to the expression of pro-inflammatory cytokines such as IL-6 and TNF-alpha [19, 20]. Additionally, IL-6 itself can activate AT1-R resulting in overexpression and production of reactive oxidative species (ROS) when RAAS is overstimulated [19]. The RAAS is also a potent oxidant stimulator, as it activates the NADH/NADPH oxidase signaling pathway, and thereby produces superoxide anions and other ROS. TNF-alpha impairs endothelial nitric oxide (NO) production in coronary arteries thereby causing vasoconstriction. Additionally, ACE plays a role in the degradation of bradykinin, which depletes NO formation as well [6, 18–20]. Overall, we have a RAAS mediated expression of ROS, inflammatory mediators, and depletion of vasodilatory NO.

This inflammation mediated cellular injury and production of ROS, activates the endothelium and increases expression of intercellular adhesion molecules (ICAM-1) and vascular cell adhesion molecules (VCAM-1), which promote endothelial damage and make cells leaky [9, 21, 22]. The endothelial damage promotes further migration of leukocytes, production of inflammatory cytokines and chemokines.

Finally, RAAS promotes thrombosis through Ang II receptors located on human platelets. Through these receptors Ang II promotes the release of thromboxane A2

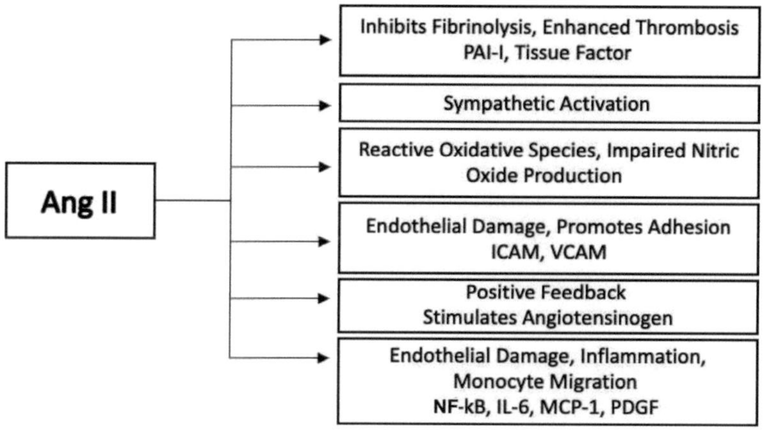

Figure 3.
Summarized effects of Ang II as it is known to cause endothelial damage, inflammation, migration and adhesion of monocytes, proliferation of vasculature and platelets and formation of atherosclerotic plaque and thrombus [6, 18–23].

and platelet derived growth factor, which promote atherosclerotic plaque formation and thrombus formation [22, 23]. Ang II involvement in endothelial dysfunction and atherosclerotic plaque formation is summarized in **Figure 3**.

4. Hypertension

Hypertension, defined as a systolic blood pressure greater than 120 and diastolic pressure greater than 80, affects a quarter of the world's population. When the etiology of hypertension is unknown, it is termed essential hypertension. When the cause of hypertension is known, by way of underlying metabolic, hormonal, neurogenic, or cardiovascular dysfunction, it is deemed as secondary hypertension [24]. As we have reviewed thus far, RAAS is responsible for maintaining sodium concentration in the blood, fluid status, and hemodynamic stability and therefore has a significant effect on blood pressure. Overactivation of RAAS can perpetuate unwanted elevations in blood pressure.

Increased levels of Ang II and subsequently aldosterone cause increases in vascular tone and hypertension. Aldosterone, a mineralocorticoid, takes its effect by binding to mineralocorticoid receptors (MR) and translocating into nucleus. Here, it integrates with cellular DNA and induces transcription of genes that regulate electrolytes and fluid balance. An over expression of aldosterone causes an elevated aldosterone-renin ratio which leads to systemic complications [4].

Patient's with primary aldosteronism (PA) and increased aldosterone levels are at higher risk for cerebrovascular complications. Although PA is not a common diagnosis, fifteen percent of patients with essential hypertension have higher than normal levels of circulating aldosterone. We can conclude that this sub-set of essential hypertension patients will have similar end-organ effects of elevated aldosterone as do patients with PA [4].

Hypertension itself can cause endovascular injury, which leads to increased production of ROS and inflammatory mediators ultimately contributing to atherosclerosis [25, 26]. The result of such endothelial injury is worsening cardiovascular disease, hypertension, and renal dysfunction. We see this manifest in the kidney with proteinuria and collagen deposition. Eventually, healthy kidney parenchyma is replaced with fibrotic tissue, leading to even more dysregulation with blood pressure homeostasis. In the cardiovascular system, inflammatory damage from overactivation of RAAS and hypertension causes calcifications and fibrosis. As such, inhibition of the RAAS system with ACE-I and ARB has become a cornerstone in therapy for hypertensive patients, particularly those with evidence of diabetes, microalbuminuria and in CAD patients overall [15, 25–28]. The details of some of the landmark clinical trials contributing to the guidelines in treatment with ACE-I and ARB are further discussed in this chapter.

5. Ischemic heart disease

Coronary artery disease (CAD) or Ischemic heart disease (IHD), develops when there is a limitation of blood flow within the coronaries. It occurs due to the gradual buildup of atherosclerotic plaque within the wall of arteries leading to reduced oxygen delivery to cardiac myocytes. It comprises a clinical spectrum based on the degree of luminal narrowing and the activation of the atherosclerotic plaque [13, 14]. The RAAS plays a vital role in the pathogenesis of CAD. Evidence supports that RAAS controls atherosclerosis through intracellular signaling pathways by mediating endothelial function, inflammation, fibrinolytic balance, growth, lipid-glucose metabolism, and its vasoconstrictor function.

Ang II has growth promoting effects by regulating growth of vascular smooth muscle cells and activating the growth associated kinase pathways. In states of ischemia, there is increased vascular endothelial growth factor (VEGF) expression. In vascular smooth muscle cells, transforming growth factor B1, platelet derived growth factor causes fibrosis and cellular hypertrophy. These angiogenic factors lead to the formation of new cells, fibrin, and collagen deposition leading to growth of the plaque and thickening of vessels [20, 21].

RAAS plays a role in altering the fibrinolytic balance as well by inhibiting fibrinolysis and enhancing thrombosis. Within the vessels, Ang II stimulates the release of plasminogen activator inhibitor - I (PAI-I) thereby reducing the fibrinolytic activity. It activates tissue factor which acts as a cofactor for factor VII, potentiating the coagulation cascade [22, 23]. The above mechanism increases the thrombogenic activity.

Ang II overexpression causes endothelial inflammation and activation of cytokine cascade thereby causing progression of atherosclerotic plaque. The silent plaque ruptures when the inflammation overwhelms the stable fibrous cap causing thrombosis and acute ischemia [13, 14].

6. Heart failure

Heart failure is a clinical syndrome categorized based on clinical signs and symptoms and further subclassified by echocardiography findings. As per the American College of Cardiology, left ventricular ejection fraction (LVEF) of ≥50% is defined as heart failure with preserved ejection fraction (HFpEF), LVEF 41-49% as heart failure with mid-range ejection fraction (HFmrEF), LVEF≤40% as heart failure with reduced ejection fraction (HFrEF). HFrEF particularly occurs after an inciting event like myocardial injury, arrhythmias, cardiomyopathies, substance abuse, infections or genetic diseases which put the heart in a state of stress leading to contractile dysfunction and cellular remodeling [29]. The circulatory changes arising from heart failure are sensed by the peripheral baroreceptors and chemoreceptors, thereby activating a sequalae of compensatory neurohormonal mechanisms. The compensatory mechanisms include activation of sympathetic nervous system (SNS) and RAAS. RAAS plays an integral role in cardiac contractility, homeostatic control of blood pressure and electrolyte-fluid balance [30, 31].

In an adult with normal circulation, the baroreceptors located in the carotid sinus and aortic arch balance the sympathetic and parasympathetic outflow from the central nervous system. Alterations in the cardiac output change the effective arterial blood volume resulting in inhibition of parasympathetic response and a reflux increase in the sympathetic vascular tone. The increased sympathetic activity leads to vasoconstriction of the renal afferent arteriole and decreases blood flow to the kidney [29, 32]. This activates renin secretion and thereby RAAS.

Renin is secreted in response to 4 main stimuli [10, 33]:

1. Decreased renal perfusion pressure sensed by baroreceptor cells in the arterial vessel wall

2. Decreased intracellular chloride levels (altered NaCl delivery)

3. Sympathetic nerve stimulation via beta-1 adrenergic receptors

4. Negative feedback by a direct action of Ang II

The pathophysiology of heart failure allows for decreased renal perfusion and increased sympathetic response, both of which cause an overactivation of the RAAS [34]. The overstimulation of RAAS in heart failure is further depicted in **Figure 4**.

In pathological states like pressure or volume overload, cardiac tissues exhibit elevated levels of renin and Ang II levels leading to cardiac hypertrophy, myocardial fibrosis, hypertensive heart disease and chronic heart failure through mechanics explained earlier. Additionally, post-infarction levels of ACE-2 have been shown to be elevated, which may explain a counter-regulatory mechanism to protect against the Ang-II mediated myocardial damage. When this natural counter-regulatory mechanism is lost in ACE-2 knockout animal models the levels of dilated cardiomyopathy were much more pronounced. Several trials have also looked at specific levels of plasma renin and HFrEF and have found that those with elevated levels had an associated worse outcome than their counterparts. In patients with advanced heart failure, baseline levels of plasma renin and plasma aldosterone are persistently high, which further exemplifies the role of RAAS in cardiac remodeling and heart failure [35–37].

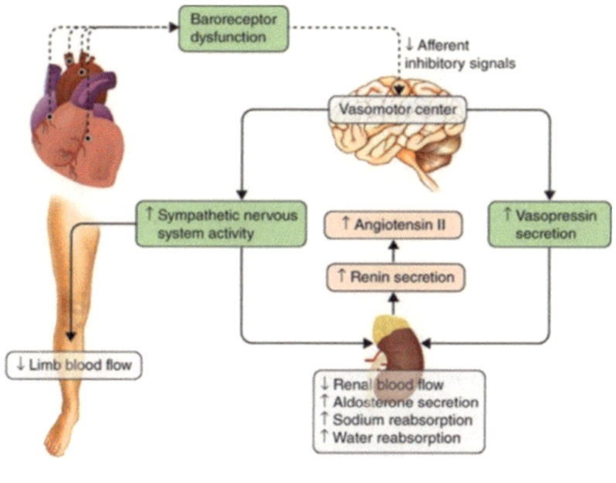

Figure 4.
The regulatory effects of RAAS as it pertains to heart failure mechanics [34]. Reproduced with permission from McGraw Hill LLC.

Innovative studies have discovered that a particular breakdown product of Ang 1-7, also known as Alamandine, has shown to prevent ventricular and vascular remodeling in animal models [11]. Studies of by-products offer areas of potential research as we grow to understand the intricacies of the molecular pathways that play a role in the development of heart failure.

7. Clinical implications

The overactivation of RAAS and its effects on the pathophysiology of hypertension, vascular stiffness, ischemia, thrombosis, and left ventricular (LV) remodeling has been well documented. As such, several medications that impede the harmful effects of the overactivation of RAAS have been shown to prevent the negative clinical outcomes. Here we review some of the landmark clinical trials that have contributed to the current guidelines and recommendations for the treatment of hypertension, ischemic heart disease and heart failure (**Table 1**).

In the treatment of hypertension, the patient's specific co-morbidities must be considered prior to initiating therapy including, race, diabetes, kidney function and other high-risk pre-existing conditions that may predispose to CV outcomes. One landmark trial, the AASK trial (2002), studied African Americans with hypertension and kidney disease and compared intensive blood pressure control versus conservative blood pressure control with ACE-I, metoprolol, and amlodipine. The two groups had no difference in the progression to CKD, however patients on ACE-I had less chronic kidney disease events and death, which solidified the use of ACE-I in patients with CKD [38].

The mainstay of treatment in patients with heart failure and CAD is blockade of the RAAS. Multiple trials highlighted in **Table 1** have been performed showing improvement in cardiovascular (CV) outcomes and reduced CV mortality.

The first trial to demonstrate improved CV outcomes with HFrEF is the CONSENSUS (1987) trial conducted among New York Heart Association (NYHA) Class IV HF and cardiomegaly patients which compared enalapril and placebo. Six-month mortality with enalapril was 26% as opposed to 44% with placebo [39]. The SOLVD (1991) treatment trial chose patients with HF and LVEF ≤35%, NYHA II-IV, with similar randomization, showing mortality reduction by 16% due to reduction of death in patients on enalapril versus placebo. This study also showed a decrease in CV related hospitalizations [40]. Further research with the V-HeFT II (1991) trial showed that ACE-I was superior in improving survival to vasodilators such as isosorbide dinitrate and hydralazine [41]. Additionally, use of ACE-I as a disease modifying drug was established post-MI in the SAVE trial (1992), which is further discussed in **Table 1** [42].

Additional studies looked to compare the effects of ACE-I versus ARB. These trials were the VALIANT (2003) trial and the OPTIMAAL (2002) trial. The VALIANT trial showed that valsartan was as effective as captopril in improving survival among patients with HF and/or LV disfunction in the post-MI period [43]. The OPTIMAAL trial compared losartan and captopril in high-risk patients after acute myocardial infarction with LV-dysfunction and heart failure and found no difference in mortality outcomes [44]. Similar studies in patients with HFpEF were conducted, including the CHARM-Preserved trial (2003) and the I-PRESERVE trial (2008). CHARM- Preserved showed that candesartan modestly reduced HF-related hospitalizations however had no effect on mortality [45]. I-PRESERVE used Irbesartan in HFpEF patients and similarly found no reduction in mortality [46].

The thought that the addition of an ARB to an ACE inhibitor could inhibit RAAS more significantly was established. This was compared in two large significant

Trial Name (Date)	Primary/ Secondary Outcomes	Inclusion Criteria	Intervention	Number of Patients / Follow-up Time	Results
AASK Trial (2002) [38]	Rate of eGFR change, progression of CKD or all-cause mortality	AA, Age 18-70, DBP >95 mmHg, HTN renal Disease, eGFR 20-65	BP control – with ramipril, amlodipine or metoprolol	1,094/4 years	No difference in progression of CKD. Use of ACE-I associated with fewer CKD events or death
CONSENSUS (1987) [39]	6 Month Mortality	NYHA IV HFrEF, optimal treatment with at least diuretic and digitalis or other medications (nitrates, prazosin, hydralazine)	Enalapril VS placebo	253/6-20 months (about 1 and a half years)	Six-month mortality with enalapril was 26% as opposed to 44% with placebo
SOLVD (1991) [40]	All-cause mortality, CV death, Death due to MI, Death due to stroke	HF, LVEF <35%, Receiving conventional therapy without ACE-I	Enalapril VS placebo	2,569/3.5 years	Enalapril reduces 4-year mortality by 16% and reduces HF hospitalizations
V-HeFT II (1991) [41]	2-year mortality, hemodynamic effects, EF, exercise tolerance, adherence	Men ages 18-75, reduced exercise tolerance, cardiac dysfunction, receiving optimal and stable therapy	Enalapril. VS ISDN/ hydralazine	804 men/2.5 years	Enalapril improved survival compared to combination of ISDN and hydralazine
SAVE (1992) [42]	All-cause mortality	Age over 21 years, MI in prior 3 days, new onset LVEF less than 40%, absence of overt signs of CHF	Captopril VS placebo	2,231/42 months (about 3 and a half years)	In patients with acute MI complicated by low EF, captopril led to 19% reduction in all-cause mortality
VALIANT (2003) [43]	All-cause mortality	Age > 18 years, Acute MI within prior 10 days complicated by HF, LVEF <35% on echocardiogram or < 40% on radionucleotide ventriculography	Valsartan VS valsartan + captopril VS captopril	14,703/ 24 months (about 2 years)	Valsartan was as effective as captopril in improving survival
OPTIMAAL (2002) [44]	All-cause mortality	50 years of age or older with confirmed acute MI and HF in acute phase or a new Q-wave anterior infarction or reinfarction	Losartan VS captopril	5,477/2.7 years	No significant change in mortality between the two drugs, however losartan was better tolerated
CHARM-Preserved (2003) [45]	Cardiovascular death or HF admission	LVEF>40%, NYHA class II-IV symptoms for at least 4 weeks, history of at least one cardiac hospitalization	Candesartan VS placebo	3,020/3 years	Candesartan modestly reduced the rate of HF-related hospitalizations. No effect on CV mortality

Trial Name (Date)	Primary/ Secondary Outcomes	Inclusion Criteria	Intervention	Number of Patients / Follow-up Time	Results
I-PRESERVE (2008) [46]	Death from any cause, hospitalization for CV disease	40 years of age or older and had NYHA class II-IV and an EF of at least 45%	Irbesartan VS placebo	4,128/49 months	Irbesartan did not improve mortality in patients with HFpEF
CHARM added (2003) [47]	CV mortality or HF hospitalizations	Age > 18 years, LVEF<40% in prior 6 months, NYHA II-IV, treatment with stable ACE-I dose for >30 days	Candesartan VS placebo	2,548/ 41 months	Addition of candesartan reduced CV mortality of HF hospitalization
VAL-HeFT (2001) [48]	All-cause mortality, cardiac arrest with resuscitation, HF hospitalization	Age > 18 years, NYHA II-IV, receipt of a fixed dose of medical therapy (ACE, digoxin, diuretics, and/or BB) for >2 weeks, EF < 40%	Valsartan VS placebo	5,010/23 months (about 2 years)	In a time where HF management included ACE but not BB, addition of ARB decreased HF hospitalizations
ON TARGET (2008) [49]	CV mortality, MI, stroke, HF hospitalization	Age over 55 years with CAD, PAD, CV disease or high-risk DM	Telmisartan VS Ramipril VS Telmisartan and Ramipril	25,620/56 months (about 4 and a half years)	Patients with CV disease or DM with complications telmisartan was as good as Ramipril in preventing death, MI, and stroke. The combination of both however had no increase in benefit and was associated with more adverse events.
RALES (1991) [50]	All-cause mortality	NYHA IV within 6 months to enrollment, NYHA III or IV at the time of enrolment, treatment with ACE and a loop diuretic, LVEF <35%	Spironolactone VS placebo	1,663/2 years	Spironolactone led to 30% reduction in all-cause mortality without significant side-effects
TOPCAT (2014) [51]	CV mortality, aborted cardiac arrest, or HF hospitalization	Age > 50 years, LVEF >45%, SBP <140 or < 160 if on 3 anti-hypertensives, serum potassium <5, elevated BNP in last 60 days, or HF hospitalization in last 12 months	Spironolactone VS placebo	3,445/3 years	Spironolactone did not reduce CV mortality however did result in a small reduction in HF hospitalizations
EMPHASIS-HF Trial (2011) [52]	CV death or hospitalization, all-cause mortality, fatal or non-fatal MI	Age > 55 years, NYHA II, EF < 30%, treatment with ACE, ARB or both, treatment with BB, CV hospitalization in last 6 months	Eplerenone VS placebo	2,737/21 months (about 2 years)	Eplerenone reduces the risk of death and hospitalization in patients with low EF and NYHA II

Trial Name (Date)	Primary/ Secondary Outcomes	Inclusion Criteria	Intervention	Number of Patients / Follow-up Time	Results
PARADIGM-HF Trial (2014) [53]	CV mortality or HF hospitalization	Age > 18 years, NYHA class II-IV, EF <35%, if no HF hospitalizations in last year BNP >150 pg./mL, ACE, or ARB and BB with stable dose, If HF hospitalization in last year BNP >100 pg./mL	ARNI VS enalapril	8,399/27 months (about 2 and a half years)	ARNI reduces CV mortality or HF hospitalizations when compared to enalapril. Also reduces all-cause mortality
PARAGON-HF Trial (2019) [54]	HF hospitalizations and CV mortality, change in NYHA class at 8 months, all-cause mortality	>50 years of age, LVEF>45%, NYHA II-IV, and at least one of the following: HF hospitalization with NT-proBNP>200 (no AFIB) or > 600 (AFIB) or NT-proBNP>300 (no AFIB) or > 900 (Afib) on screening visit ECG	ARNI VS valsartan alone	4,822/35 months (about 3 years)	ARNI did not lower hospitalizations or death from CV causes, however there was a modest improvement in NYHA class and a slower decline in renal function than what was seen in valsartan alone
PIONEER-HF Trial (2019) [55]	Time-averaged change in NT-proBNP concentration from baseline through weeks 4-8	Age > 18 years, LVEF<40%, NT-proBNP of 1600 pg./mL or more, or BNP of 400 pg./mL or more, receiving diagnosis of acute decompensated HF up to10 days after presentation	ARNI versus enalapril	881/2 years	ARNI decreased NT-proBNP compared to enalapril therapy without significant change in rate of adverse events
ALTITUDE Trial (2012) [56]	Death from CV causes, nonfatal MI, nonfatal stroke, ESRD, death attributable to kidney failure, or the need for RRT	35 years or older with type 2 diabetes and evidence of microalbuminuria, macroalbuminuria, or cardiovascular disease	Aliskiren VS placebo	8,561/32 months (about 2 and a half years)	The addition of aliskiren to standard therapy in patients with type 2 diabetes who are at elevated risk for CV and renal events is potentially harmful

Summarized landmark clinical trials depicting the benefits of RAAS inhibition in cardiac and renal patients. Abbreviations: CKD: Chronic kidney disease, AA: African American, SBP: systolic blood pressure, DBP: diastolic blood pressure, CV: cardiovascular, EF: ejection fraction, LV: left ventricular, MI: Myocardial infarct, ESRD: End stage renal disease, RRT: renal replacement therapy.

Table 1.
Trials documenting improvement in cardiovascular outcomes and reduced cardiovascular mortality with renin-angiotensin-aldosterone system inhibition.

trials. The CHARM-added trial compared symptomatic HF patients with LVEF ≤40% who were already on an ACE inhibitor with either addition of candesartan or placebo. This trial showed a reduction in CV mortality and HF hospitalizations; however, it was accompanied by a significant increase in hyperkalemic events [47]. The Val-HeFT (2001) compared patients with symptomatic HF, LVEF <40% with LV dilatation and on ACE inhibitors by adding either valsartan or placebo. There was no effect on mortality however, there was a 23% reduction in HF hospitalization in the treatment group [48]. Finally, the ONTARGET trial (2008) compared ramipril to telmisartan to a combination of both in patients with CV disease or diabetes with complications and found that the combination of telmisartan plus ramipril had no increase in benefit and was associated with more adverse events [49].

Several trials looking at the effects of aldosterone antagonists and heart failure patients were conducted with overall favorable results. Patients benefit from reduced sympathetic stimulation and alleviate fluid overload from sodium and water retention through aldosterone blockade. The RALES trial (1999) studied the role of spironolactone in patients with LVEF≤35% and NYHA class III-IV, which showed that Spironolactone, along with ACE-I (as most patients were already on ACE-I) showed a 11% reduction in CV mortality compared to placebo [50]. The TOPCAT trial (2014) done in patients with HFpEF and controlled blood pressures to receive spironolactone or placebo. This study conversely showed that spironolactone did not reduce CV mortality however did result in a small reduction in HF hospitalizations [51]. Another trial, the EMPHASIS-HF trial (2011), looked at Eplerenone versus placebo in HF patients, NYHA class II, showed that Eplerenone reduced the risk of death and hospitalizations in patients with HF [52].

A newer group of RAAS inhibition medications combining an ARB and neprilysin inhibitor (ARNI) was studied in 2014 in the PARADIGM-HF trial. Neprilysins are key enzymes in the degradation of natriuretic peptides. They increase endogenous natriuretic peptide levels including bradykinin, thereby promoting vasodilation and natriuresis. Neprilysins were initially attempted with an ACE inhibitor combination however this led to incidences of angioedema given increased levels of bradykinin. PARADIGM - HF trial was conducted in patients with symptomatic HF and LVEF ≤40% assigned to enalapril alone or valsartan-sacubitril combination. This showed significant reduction in CV mortality, all-cause mortality, and HF hospitalizations with no increase in angioedema events [53]. The PARAGON-HF trial (2019) studied ARNI versus valsartan alone in HFpEF patients with EF > 45% and NYHA II to IV and showed that ARNI did not lower hospitalizations or death from CV causes, however there was a modest improvement in NYHA class and a slower decline in renal function than what was seen in valsartan alone [54]. The PIONEER-HF trial (2019) showed that initiated of ARNI versus enalapril in acute diastolic heart failure patients allowed for significant reductions in HF biomarker, NT-proBNP, without significant change in adverse effects [55].

Direct renin inhibitors have been attempted with the goal of reducing renin and thereby the entire RAAS cascade. The ALTITUDE trial (2012) added aliskiren to patients with diabetes type 2 in order to prevent kidney disease and CV outcomes. These patients were on ACE-I however the addition of aliskiren led to an increase in CV mortality, hypotension, and adverse hyperkalemic events. The trial was stopped early due to higher mortality findings [56].

8. Summary and conclusions

RAAS is a complex and evolving pathway that has been implicated in the pathogenesis of endothelial damage, atherosclerosis, and cardiac remodeling. Inhibition of the negative effects of overactivated RAAS has shown to cause morbidity and mortality benefits in cardiovascular disease outcomes. Significant research has yet to be performed on the possibility of stimulating the counter-regulatory effects of RAAS through AT2-R and MAS-R. Such mechanisms are still being studied in animal models; however, the effects of AT2-R and MAS-R offer potential areas of continued research and potential targets for future therapy.

Conflict of interest

No conflicts of interest exist for this work by any of the authors.

Author details

Violeta Capric[1], Harshith Priyan Chandrakumar[1], Jessica Celenza-Salvatore[1] and Amgad N. Makaryus[2,3*]

1 Department of Medicine, State University of New York-Downstate, Downstate-Health Science University, Brooklyn, NY, United States

2 Department of Cardiology, Donald and Barbara Zucker School of Medicine at Hofstra/Northwell, United States

3 Department of Cardiology, Nassau University Medical Center, East Meadow, NY, United States

*Address all correspondence to: amakaryu@numc.edu

References

[1] WHO's Global Health Estimates (2020, December 9). The top 10 causes of death. Retrieved December 24, 2020, from https://www.who.int/news-room/fact-sheets/detail/the-top-10-causes-of-death

[2] Ferrario, C. M. Role of Angiotensin II in Cardiovascular Disease Therapeutic Implications of More than a Century of Research. *J. Renin Angiotensin Aldosterone Syst.* 2006;7: 3-14

[3] Patel, S., Rauf, A., Khan, H., & Abu-Izneid, T. (2017). Renin-angiotensin-aldosterone (RAAS): The ubiquitous system for homeostasis and pathologies. Biomedicine & Pharmacotherapy, 94, 317-325. doi:10.1016/j.biopha.2017.07.091

[4] Muñoz-Durango, N., Fuentes, C., Castillo, A., González-Gómez, L., Vecchiola, A., Fardella, C., & Kalergis, A. (2016). Role of the Renin-Angiotensin-Aldosterone System beyond Blood Pressure Regulation: Molecular and Cellular Mechanisms Involved in End-Organ Damage during Arterial Hypertension. International Journal of Molecular Sciences, 17(7), 797. doi:10.3390/ijms17070797

[5] Paz Ocaranza M, Riquelme JA, García L, Jalil JE, Chiong M, Santos RAS, Lavandero S. Counter-regulatory renin-angiotensin system in cardiovascular disease. *Nat Rev Cardiol.* 2020 Feb;17(2):116-129.

[6] Jacoby, D.S., Rader, D.J. Renin Angiotensin System and Atherothrombotic Disease: From Genes to Treatment. *Arch. Intern Med.* 2003;163: 1155-1164

[7] Slagman MC, Navis G, Laverman GD. Dual blockade of the renin-angiotensin-aldosterone system in cardiac and renal disease. *Curr Opin Nephrol Hypertens.* 2010 Mar;19(2):140-52.

[8] Agrawal V, Gupta JK, Qureshi SS, Vishwakarma VK. Role of cardiac renin angiotensin system in ischemia reperfusion injury and preconditioning of heart. *Indian Heart J.* 2016;68(6):856-861. doi:10.1016/j.ihj.2016.06.010

[9] Paul M, Poyan Mehr A, Kreutz R. Physiology of local renin-angiotensin systems. *Physiol Rev. 2006 Jul;86(3): 747-803. doi: 10.1152/physrev.00036.2005. PMID: 16816138.*

[10] Atlas SA. The renin-angiotensin aldosterone system: pathophysiological role and pharmacologic inhibition. J Manag Care Pharm. 2007 Oct;13(8 Suppl B):9-20. doi: 10.18553/jmcp.2007.13.s8-b.9. PMID: 17970613

[11] Qaradakhi, T., Apostolopoulos, V., and Zulli, A. (2016). Angiotensin (1-7) and alamandine: similarities and differences. *Pharmacol. Res.* 111, 820-826. doi: 10.1016/j.phrs.2016.07.025

[12] Montezano A. C., Nguyen D. C. A., Rios F. J., Touyz R. M. Angiotensin II and vascular injury. *Curr. Hypertens. Rep.* 2014; 16:431

[13] Silva GM, França-Falcão MS, Calzerra NTM, et al. Role of Renin-Angiotensin System Components in Atherosclerosis: Focus on Ang-II, ACE2, and Ang-1-7. *Front Physiol*. 2020;11:1067.

[14] De Mello WC, Danser AH. Angiotensin II and the heart : on the intracrine renin-angiotensin system. *Hypertension.* 2000 Jun;35(6):1183-8.

[15] Dandona P, Dhindsa S, Ghanim H, Chaudhuri A. Angiotensin II and inflammation: the effect of angiotensin-converting enzyme inhibition and angiotensin II receptor blockade. *J Hum Hypertens.* 2007 Jan;21(1):20-7.

[16] Ludwig,A., Wilck, N. Targeting the Ubiquitin-Proteasome System in

Atherosclerosis: Status Quo, Challenges, and Perspectives. *Antioxidants & Redox Signaling. 2014;*21(17).

[17] Pueyo ME, Gonzalez W, Nicoletti A, Savoie F, Arnal JF, Michel JB. Angiotensin II stimulates endothelial vascular cell adhesion molecule-1 via nuclear factor-kappaB activation induced by intracellular oxidative stress. *Arterioscler Thromb Vasc Biol*. 2000 Mar;20(3):645-51.

[18] Guanghong Jia, Annayya R. Aroor, Michael A. Hill, James R. Sowers. Role of Renin-Angiotensin-Aldosterone System Activation in Promoting Cardiovascular Fibrosis and Stiffness. *Hypertension*. 2018 ;72:537-548.

[19] Rajagopalan S, Kurz S, Münzel T, et al. Angiotensin II-mediated hypertension in the rat increases vascular superoxide production via membrane NADH/NADPH oxidase activation. Contribution to alterations of vasomotor tone. *J Clin Invest.* 1996;97(8):1916-1923.

[20] Wassmann S, Nickenig G. Pathophysiological regulation of the AT1-receptor and implications for vascular disease. *J Hypertens Suppl.* 2006 Mar;24(1):S15-21.

[21] Terry KW Ma, Kevin KH Kam, Bryan P Yan and Yat-Yin Lam. Renin–angiotensin–aldosterone system blockade for cardiovascular diseases: current status. British Journal of Pharmacology (2010), 160, 1273-1292

[22] Soejima H, Ogawa H, Yasue H, Kaikita K, Takazoe K, Nishiyama K, Misumi K, Miyamoto S, Yoshimura M, Kugiyama K, Nakamura S, Tsuji I. Angiotensin-converting enzyme inhibition reduces monocyte chemoattractant protein-1 and tissue factor levels in patients with myocardial infarction. *J Am Coll Cardiol.* 1999 Oct;34(4):983-8.

[23] Vaughan DE. Angiotensin and vascular fibrinolytic balance. *Am J Hypertens.* 2002 Jan;15(1 Pt 2):3S-8S.

[24] Whelton, Paul K., et al. 2017 ACC/AHA/AAPA/ABC/ACPM/AGS/APhA/ASH/ASPC/NMA/PCNA Guideline for the Prevention, Detection, Evaluation, and Management of High Blood Pressure in Adults: A Report of the American College of Cardiology/American Heart Association Task Force on Clinical Practice Guidelines. *Journal of the American College of Cardiology*, 1 May 2018, www.jacc.org/doi/10.1016/j.jacc.2017.11.006?_ga=2.16409077.234551544.1610300589-914503138.1610300589&.

[25] Olafiranye, O., Zizi, F., Brimah, P., Jean-Louis, G.,Makaryus, A. N., McFarlane, S., & Ogedegbe, G. Management of hypertension among patients with coronary heart disease. *International Journal of Hypertension.* 2011;2(1), 34-41. doi:10.4061/2011/653903

[26] Maraj, Ilir, et al. "Hypertension Management in the High Cardiovascular Risk Population." *International Journal of Hypertension*, Hindawi (2013) 2013: 1-7. https://doi.org/10.1155/2013/382802

[27] Fridman, David, et al. "Diabetes and Coronary Artery Disease – Pathophysiologic Insights and Therapeutic Implications." *IntechOpen*, IntechOpen, 18 Nov. 2015, dx.doi.org/10.5772/61875.

[28] Makaryus AN, Akhrass P, McFarlane SI. Treatment of hypertension in metabolic syndrome: implications of recent clinical trials. *Curr Diab Rep* (2009) 9(3):229-37.

[29] Yancy CW, Jessup M, Bozkurt B, Butler J, Casey DE Jr, Colvin MM, Drazner MH, Filippatos GS, Fonarow GC, Givertz MM, Hollenberg SM, Lindenfeld J, Masoudi FA, McBride PE, Peterson PN,

Stevenson LW, Westlake C. 2017 ACC/AHA/HFSA Focused Update of the 2013 ACCF/AHA Guideline for the Management of Heart Failure: A Report of the American College of Cardiology/American Heart Association Task Force on Clinical Practice Guidelines and the Heart Failure Society of America. *Circulation. 2017 Aug 8;136(6):e137-e161. doi: 10.1161/CIR.0000000000000509. Epub 2017 Apr 28. PMID: 28455343*

[30] Dostal DE, Baker KM. The cardiac renin-angiotensin system: conceptual, or a regulator of cardiac function? *Circ Res.* 1999 Oct 1;85(7):643-50. doi: 10.1161/01.res.85.7.643. PMID: 10506489

[31] Ruzicka M, Leenen FH. Relevance of blockade of cardiac and circulatory angiotensin-converting enzyme for the prevention of volume overload-induced cardiac hypertrophy. *Circulation*. 1995 Jan 1;91(1):16-9. doi: 10.1161/01.cir.91.1.16. PMID: 7805197.

[32] Hartupee J, Mann DL. Neurohormonal activation in heart failure with reduced ejection fraction. *Nat Rev Cardiol.* 2017 Jan;14(1):30-38. doi: 10.1038/nrcardio.2016.163. Epub 2016 Oct 6. PMID: 27708278; PMCID: PMC5286912.

[33] Verbrugge FH, Tang WH, Mullens W. Renin-Angiotensin-aldosterone system activation during decongestion in acute heart failure: friend or foe? JACC Heart Fail. 2015 Feb;3(2):108-11. doi: 10.1016/j.jchf.2014.10.005. Epub 2014 Oct 31. PMID: 25543974

[34] Mann DL, Chakinala M. Heart Failure: Pathophysiology and Diagnosis. In: Jameson J, Fauci AS, Kasper DL, Hauser SL, Longo DL, Loscalzo J. eds. *Harrison's Principles of Internal Medicine, 20e*. McGraw-Hill; Accessed December 22, 2020.

[35] Francis GS, Benedict C, Johnstone DE, Kirlin PC, Nicklas J, Liang CS, Kubo SH, Rudin-Toretsky E, Yusuf S. Comparison of neuroendocrine activation in patients with left ventricular dysfunction with and without congestive heart failure. A substudy of the Studies of Left Ventricular Dysfunction (SOLVD). *Circulation. 1990 Nov;82(5):1724-9. doi: 10.1161/01.cir.82.5.1724. PMID: 2146040.*

[36] Kostin S, Hein S, Arnon E, Scholz D, Schaper J. The cytoskeleton and related proteins in the human failing heart. *Heart Fail Rev.* 2000 Oct;5(3):271-80. doi: 10.1023/A:1009813621103. PMID: 16228910.

[37] Mentz RJ, Stevens SR, DeVore AD, Lala A, Vader JM, AbouEzzeddine OF, Khazanie P, Redfield MM, Stevenson LW, O'Connor CM, Goldsmith SR, Bart BA, Anstrom KJ, Hernandez AF, Braunwald E, Felker GM. Decongestion strategies and renin-angiotensin-aldosterone system activation in acute heart failure. *JACC Heart Fail.* 2015 Feb;3(2):97-107. doi: 10.1016/j.jchf.2014.09.003. Epub 2014 Oct 31. PMID: 25543972; PMCID: PMC4324057.

[38] Wright JT Jr, Bakris G, Greene T, Agodoa LY, Appel LJ, Charleston J, Cheek D, Douglas-Baltimore JG, Gassman J, Glassock R, Hebert L, Jamerson K, Lewis J, Phillips RA, Toto RD, Middleton JP, Rostand SG; African American Study of Kidney Disease and Hypertension Study Group. Effect of blood pressure lowering and antihypertensive drug class on progression of hypertensive kidney disease: results from the AASK trial. *JAMA*. 2002 Nov 20;288(19):2421-31. Erratum in: *JAMA*. 2006 Jun 21;295(23):2726.

[39] CONSENSUS Trial Study Group. Effects of enalapril on mortality in severe congestive heart failure. Results of the Cooperative North Scandinavian Enalapril Survival Study (CONSENSUS). *N Engl J Med*. (1987) Jun 4;316(23):1429-35.

[40] SOLVD Investigators, Yusuf S, Pitt B, Davis CE, Hood WB, Cohn JN. Effect of enalapril on survival in patients with reduced left ventricular ejection fractions and congestive heart failure. *N Engl J Med*. (1991) 325(5):293-302.

[41] Cohn JN, Johnson G, Ziesche S, Cobb F, Francis G, Tristani F, Smith R, Dunkman WB, Loeb H, Wong M, et al. A comparison of enalapril with hydralazine-isosorbide dinitrate in the treatment of chronic congestive heart failure. *N Engl J Med. (*1991) 325(5):303-10.

[42] Pfeffer MA, Braunwald E, Moyé LA, Basta L, Brown EJ Jr, Cuddy TE, Davis BR, Geltman EM, Goldman S, Flaker GC, et al. Effect of captopril on mortality and morbidity in patients with left ventricular dysfunction after myocardial infarction. Results of the survival and ventricular enlargement trial. The SAVE Investigators. N Engl J Med. 1992 Sep 3;327(10):669-77. doi: 10.1056/NEJM199209033271001. PMID: 1386652.

[43] M. A. Pfeffer, J. J. McMurray, E. J. Velazquez et al., Valsartan in acute myocardial infarction trial investigators. Valsartan, captopril, or both in myocardial infarction complicated by heart failure, left ventricular dysfunction, or both. *The New England Journal of Medicine* (2003) 349:1893-1906

[44] Dickstein K, Kjekshus J. the OPTIMAAL Trial Steering Committee, for the OPTIMAAL Study Group. Effects of losartan and captopril on mortality and morbidity in high-risk patients after acute myocardial infarction: the OPTIMAAL randomized trial. *Lancet.* 2002; 360: 752-760

[45] Yusuf S, Pfeffer MA, Swedberg K, Granger CB, Held P, McMurray JJ, Michelson EL, Olofsson B, Ostergren J; CHARM Investigators and Committees. Effects of candesartan in patients with chronic heart failure and preserved left-ventricular ejection fraction: the CHARM-Preserved Trial. *Lancet*. (2003) 362(9386):777-81.

[46] Massie BM, Carson PE, McMurray JJ, Komajda M, McKelvie R, Zile MR, Anderson S, Donovan M, Iverson E, Staiger C, Ptaszynska A; I-PRESERVE Investigators. Irbesartan in patients with heart failure and preserved ejection fraction. *N Engl J Med*. (2008) 359(23):2456-67.

[47] McMurray JJ, Ostergren J, Swedberg K, Granger CB, Held P, Michelson EL, Olofsson B, Yusuf S, Pfeffer MA; CHARM Investigators and Committees. Effects of candesartan in patients with chronic heart failure and reduced left-ventricular systolic function taking angiotensin-converting-enzyme inhibitors: the CHARM-Added trial. Lancet. (2003) 362(9386):767-71.

[48] Cohn JN, Tognoni G; Valsartan Heart Failure Trial Investigators. A randomized trial of the angiotensin-receptor blocker valsartan in chronic heart failure. *N Engl J Med*. (2001) 345(23):1667-75.

[49] ONTARGET Investigators: Yusuf S, Teo KK, et al.: Telmisartan, ramipril, or both in patients at high risk for vascular events. *N Engl J Med* (2008) 358:1547-1559.

[50] Pitt B, Zannad F, Remme WJ, Cody R, Castaigne A, Perez A, Palensky J, Wittes J. The effect of spironolactone on morbidity and mortality in patients with severe heart failure. Randomized Aldactone Evaluation Study Investigators. *N Engl J Med.* (1999) 341(10):709-17.

[51] Pitt B, Pfeffer MA, Assmann SF, Boineau R, Anand IS, Claggett B, Clausell N, Desai AS, Diaz R, Fleg JL, Gordeev I, Harty B, Heitner JF, Kenwood CT, Lewis EF, O'Meara E,

Probstfield JL, Shaburishvili T, Shah SJ, Solomon SD, Sweitzer NK, Yang S, McKinlay SM; TOPCAT Investigators. Spironolactone for heart failure with preserved ejection fraction. *N Engl J Med*. (2014) 370(15):1383-92.

[52] Zannad, F., McMurray, J. J. V., Krum, H., van Veldhuisen, D. J., Swedberg, K., Shi, H., ... Pitt, B. Eplerenone in Patients with Systolic Heart Failure and Mild Symptoms. *New England Journal of Medicine* (2011) 364(1), 11-21

[53] McMurray JJ, Packer M, Desai AS, Gong J, Lefkowitz MP, Rizkala AR, Rouleau JL, Shi VC, Solomon SD, Swedberg K, Zile MR; PARADIGM-HF Investigators and Committees. Angiotensin-neprilysin inhibition versus enalapril in heart failure. *N Engl J Med*. (2014) 371(11):993-1004.

[54] Solomon, Scott D., et al. Angiotensin–Neprilysin Inhibition in Heart Failure with Preserved Ejection Fraction: NEJM. *N Engl J Med* (2019) 381:1609-1620

[55] Velazquez, Eric J., et al. Angiotensin–Neprilysin Inhibition in Acute Decompensated Heart Failure: NEJM. *N Engl J Med* (2019) 380:539-548

[56] Parving HH, Brenner BM, McMurray JJ, de Zeeuw D, Haffner SM, Solomon SD, Chaturvedi N, Persson F, Desai AS, Nicolaides M, Richard A, Xiang Z, Brunel P, Pfeffer MA; ALTITUDE Investigators. Cardiorenal end points in a trial of aliskiren for type 2 diabetes. N Engl J Med. 2012 Dec 6;367(23):2204-13.

Chapter 2

Diabetes and Renin-Angiotensin-Aldosterone System: Pathophysiology and Genetics

A.H.M. Nurun Nabi and Akio Ebihara

Abstract

Diabetes mellitus (DM) is a metabolic disorder and characterized by hyperglycemia. Being a concern of both the developed and developing world, diabetes is a global health burden and is a major cause of mortality world-wide. The most common is the type 2 diabetes mellitus (T2DM), which is mainly caused by resistance to insulin. Long-term complications of diabetes cause microvascular related problems (eg. nephropathy, neuropathy and retinopathy) along with macrovascular complications (eg. cardiovascular diseases, ischemic heart disease, peripheral vascular disease). Renin-angiotensin-aldosterone system (RAAS) regulates homeostasis of body fluid that in turn, maintains blood pressure. Thus, RAAS plays pivotal role in the pathogenesis of long-term DM complications like cardiovascular diseases and chronic kidney diseases. T2DM is a polygenic disease, and the roles of RAAS components in insulin signaling pathway and insulin resistance have been well documented. Hyperglycemia has been found to be associated with the increased plasma renin activity, arterial pressure and renal vascular resistance. Several studies have reported involvement of single variants within particular genes in initiation and development of T2D using different approaches. This chapter aims to investigate and discuss potential genetic polymorphisms underlying T2D identified through candidate gene studies, genetic linkage studies, genome wide association studies.

Keywords: diabetes, type 2 diabetes, renin-angiotensin-aldosterone system, hypertension, gene polymorphism, genome wide association study, genetics, COVID-19

1. Introduction

Diabetes is a global health burden and one of the leading causes of morbidity world-wide [1]. Diabetes mellitus (DM) is a metabolic disorder characterized by polydipsia, polyphagia, polyurea and weight loss due to hyperglycemia, which means persistent elevated levels of plasma glucose. The prolonged hyperglycemia results in long-term impediments of diabetes that cause macrovascular complications including cardiovascular diseases (CVDs) and other vascular complications including nephropathy (end-stage renal disease) or retinopathy (leading to blindness) [2]. On the other hand, renin-angiotensin-aldosterone system (RAAS) plays an important role in maintaining blood pressure and body fluid [3]. Inappropriate activation of RAAS contributes to the hemodynamic abnormalities that lead to endothelial dysfunction, hypertension, and CVD [3, 4].

Diabetes, hypertension and CVDs, are important risk factors for severity and mortality in people infected with coronavirus infectious disease 2019 (COVID-19) [5, 6]. Both Type 2 diabetes (T2D), the commonest form of diabetes and hypertension are multifactorial and polygenic diseases caused by the association of both genetic and environmental factors. Understanding the underlying genetic causes of susceptibility to these diseases is important for people's health and health-related quality of life worldwide. In this chapter, we describe the pathophysiology of T2D and RAAS and their associated risks analyzed in term of genetic variants.

2. Diabetes

Diabetes is a global epidemic affecting people of both the developed and developing world. According to International Diabetes Federation, 9.3% of the world population had diabetes in 2019 and predicted that by 2045 about 10.9% of the world population may suffer from diabetes [7]. Prevalence of diabetes is increasing both in developing and developed countries. About 79% of the diabetic patients live in low-income or lower middle-income countries of which more than 60% belongs to Asian countries while rest of them are habitant of developed world [8]. Notably, diabetes is a health concern in adults compare to other age groups and it has been projected that between the years 2010 to 2030, developing countries will harbor 69% more adults with diabetes while 20% more adults with diabetes will be residing in developed countries [9]. Persistent elevated levels of plasma glucose result in long-term impediments of diabetes that cause macrovascular complications including CVDs, peripheral vascular disease, stroke and microvascular complications including nephropathy that leads to end-stage renal disease, retinopathy leading to blindness, neuropathy that causes damage to the nerves [2].

Diabetes can be classified into the following types [10]:

i. Type 1 diabetes (T1DM; due to autoimmune β-cell destruction, usually leading to absolute insulin deficiency, including latent autoimmune diabetes of adulthood).

ii. Type 2 diabetes (T2DM; due to a progressive loss of adequate β-cell insulin secretion frequently on the background of insulin resistance).

iii. Gestational diabetes mellitus (diabetes diagnosed in the second or third trimester of pregnancy that was not clearly overt diabetes prior to gestation).

iv. Specific types of diabetes due to other causes, eg. monogenic diabetes syndromes (such as neonatal diabetes and maturity-onset diabetes of the young), diseases of the exocrine pancreas (such as cystic fibrosis and pancreatitis), and drug- or chemical-induced diabetes (such as with glucocorticoid use, in the Human Immunodeficiency Virus treatment, or after organ transplantation).

Of the major types, T2DM is the commonest form. T2D was caused by developing insulin resistance due to lifestyle, obesity, reduced physical activity [3]. Individuals with T2DM will have seven to ten years shorter life span compare to non-diabetic individuals and 80% patients with T2DM develop cardiovascular disease [11]. CVD like coronary artery disease is responsible for the 2–4 fold increased rate of death in adults [12, 13]. Diabetes being considered as the independent risk factor from other such factors as age, gender, smoking, weight for dying from liver disease, lung disease, cancer, mental disorders, cardiovascular complications [14]. Moreover, people are more prone to infections or infectious diseases who have already developed diabetes [15] due to high levels of glucose in blood that favors immune dysfunction by modulating both innate (alteration of neutrophil functions) and adaptive (reducing T cell response) immune response [16–20]. Most recent incidence of pandemic has revealed that the severity of COVID-19 exaggerates in individuals with hyperglycemia due to augmented production of pro-inflammatory cytokines as well as poor innate immunity [21]. The severe acute respiratory syndrome coronavirus 2 (SARS-CoV-2) infection severely affects the survival rate of the infected individuals [21] with diabetes as it is a critical comorbidity [22].

T2D is a multifactorial and polygenic diseases caused by the association of different risk alleles located on multiple genes. Environmental factors modulating gene–gene interaction and/or expression are believed to be contributing factor for the development of T2D. Thus, genetic variants associated with T2DM are not only important for prediction and prevention of the disease along with its associated complications, but also will facilitate early treatment as well as need-based bona fide management of the disease.

3. Renin-angiotensin-aldosterone system

RAAS is one of the multifaced systems, which maintains homeostasis of body fluids, electrolyte balance and thus, regulates blood pressure [3, 23, 24]. Renin, initially known as pressor hormone, is an aspartic protease and it's only known substrate is angiotensinogen (AGT) [25]. Angiotensin converting enzyme (ACE) is a peptidase that is mainly found in the capillaries of lung followed by endothelial and kidney epithelial cells in human [26]. The classical RAAS involves cleavage of AGT for release of a small decapeptide, angiotensin-I (Ang-I). The peptidase ACE then converts Ang-I into an octapeptide, angiotensin-II (Ang-II). RAAS activity is intrinsically high in the lung where ACE level is very high and thus, a major site of systemic Ang-II synthesis.

The Ang-II is the most potent hormone peptide that utilizes G-protein coupled receptors (GPCRs) called angiotensin type 1 and type 2 receptors (AT1R and AT2R) to mediate physiological functions. Ang-II facilitates vasoconstriction, cell proliferation, cell hypertrophy, anti-natriuresis, fibrosis, and atherosclerosis using AT1R [27] while, via AT2R, the peptide elicits vasodilation, anti-proliferation, anti-hypertrophy, anti-fibrosis, anti-thrombosis, and anti-angiogenesis [28] (**Figure 1**). Ang-II also stimulates the production of the steroid hormone, aldosterone, which is the final product of the RAAS cascade. Aldosterone binds to the mineralocorticoid receptor and regulates the transcription of target genes, resulting in the upregulation of electrolyte flux pathways in the kidney. Dysregulation of RAAS can lead to adverse effects on fluid homeostasis, which in turn may lead to organ damage followed by CVDs.

Angiotensin converting enzyme 2 (ACE2) is a homolog of ACE. ACE2 is also highly expressed in the lung. The main activity of ACE2 is to degrade Ang-II into angiotensin 1–7 (Ang 1–7) by hydrolyzing of the C-terminal residue [29]. Thus, ACE2, in the lung, have a role in adjusting the balance of circulating Ang-II/Ang 1–7 levels. Also, product of ACE2 facilitates vasodilation and therefore opposing the role of ACE product (i.e. Ang-II). Ang 1–7 is expected to exert its action through the MAS-related (MAS1) GPCR [30]. It is evident that insulin exhibits adverse effects

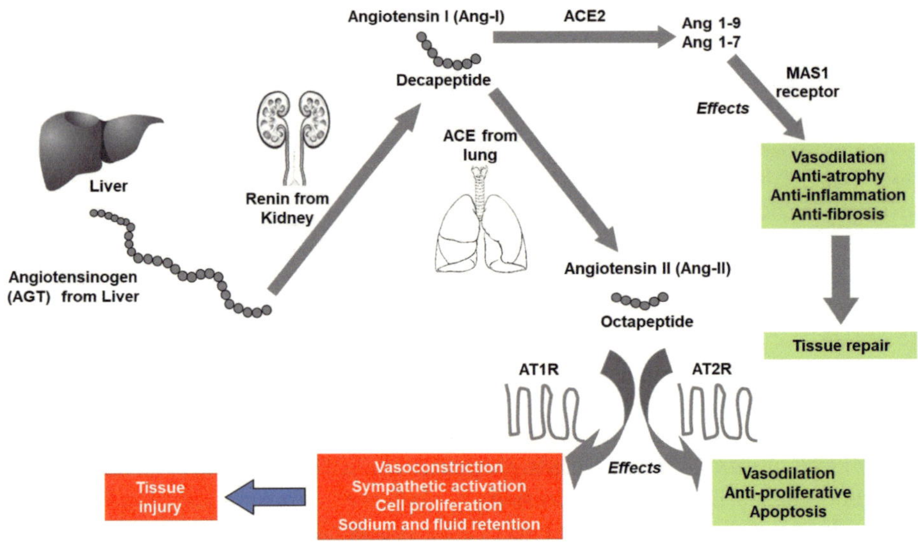

Figure 1.
Renin-angiotensin-aldosterone system (RAAS) and its linkage to type 2 diabetes mellitus (T2DM). The classical RAAS shows angiotensin-II (Ang-II) dependent pathway mediated different physiological effects via G-protein coupled receptors (GPCR) called angiotensin type 1 and type 2 receptors (AT1R and AT2R). Renin, secreted from kidney, regulates the rate limiting step of this pathway by converting its liver originated substrate angiotensinogen (AGT) into a decapeptide, angiotensin-I (Ang-I). Ang-I is converted into an octapeptide Ang-II by angiotensin converting enzyme (ACE). Ang-II binds to AT1R and AT2R to mediate the counterbalanced physiological functions. Angiotensin converting enzyme 2 (ACE2) is to cleave Ang-II into angiotensin 1–7 (Ang 1–7), which exerts the vasodilation effects.

on the structural and functional features of islet cells by inducing Ang-II mediated oxidative stress [31]. Through AT1R, Ang-II inhibits course of insulin action in vascular and skeletal muscle tissue, interferes insulin signaling via phosphatidylinositol 3-kinase and its downstream protein kinase B (Akt) signaling pathway [32].

Increased vasoconstriction and renal sodium reabsorption along with enhanced secretion of aldosterone results overactivation of RAAS followed by metabolic modulation leading to altered blood pressure and development of insulin resistance [33, 34]. Aldosterone has the ability to impair insulin signaling pathway by down-regulating insulin receptor substrate-1 (IRS-1) in vascular smooth muscle cells [35] and thus, contributes to the development of and/or deteriorating metabolic disorders including disruption of glucose homeostasis [36].

The (pro)renin receptor [(P)RR], cloned almost two decades before in 2002 [37], has now been considered as one of the pivotal members of RAAS. Modulation of renin/prorenin takes place after binding to their receptor. After binding to (P)RR, the enzymatic activity of renin increases while the proactive form of renin known as prorenin gets activated non-proteolytically and exhibits renin activity [38, 39]. Binding to (P)RR with prorenin causes a change in conformation within the prosegment region followed by opening of the active site and making it accessible to the substrate, AGT [39, 40]. Thus, receptor mediated activity of renin and prorenin possibly activate tissue specific renin-angiotensin system in an Ang-II dependent manner, which ultimately could contribute in modulating local RAAS. (P)RR has been found to be ubiquitously expressed in brain, heart, placenta, liver, pancreas and kidney [37]. The association between *(P)RR* gene polymorphism and high blood pressure has been demonstrated in Caucasian and Japanese male subjects [41, 42]. In another study with transgenic rats over expressing (P)RR in smooth muscles it was reported to elevate blood pressure and increase heart rate in their models [43]. A single mutation in exon 4 of *(P)RR* gene is associated with mental retardation and epilepsy [44] while a silent mutation in exon 4 on human (P)RR facilitates enhanced expression of c321C > T that lacked exon 4 [44]. Though presence of this single nucleotide polymorphism (SNP) does not bring any change as far as the renin binding ability is concerned but it modulates ERK1/2 activation [44], which may in turn modifies gene expression pattern.

RAAS mediates diverse functions by the action of angiotensin receptors (**Figure 1**) and has the link to cancer through tissue remodeling, inflammation, angiogenesis and apoptosis [45, 46]. Genetic and epidemiological studies showed that polymorphism of the RAS components contribute to the risk of cancer. Either the insertion/deletion (I/D) polymorphisms of *ACE* or *AGT* M235T SNP confer the risks for developing breast cancer [45]. Two *AT1R* SNPs are associated with risk for renal cell cancer, and its associations are stronger in subjects with hypertension [47]. Although the identified SNPs could be a marker of disease linked to another disease-causing SNP, rather than the disease-causing SNP itself [47], further studies are warranted to clarify cancer etiology involving the RAS components.

4. Diabetes and RAAS

Development of insulin resistance at the cellular level is initiated by Ang-II and aldosterone via increasing oxidative stress and altering insulin signaling (**Figure 2**). Ang-II is also responsible for generating pancreatic β-cell oriented oxidative stress, inflammation, and apoptosis. Evidence also suggested involvement of aldosterone in diminished glucose induced insulin secretion from pancreas [33].

The therapeutic approaches for lowering glucose levels significantly reduces the chance of developing diabetes associated microvascular complications while modest improvement has been observed in case of improving diabetes associated macrovascular complications [48, 49]. A case–control study conducted in German population demonstrated increased prevalence of T2D among individuals with hypertension and higher concentration of aldosterone (but low Ang-II level and low plasma renin activity) compared to the control hypertensive individuals [50].

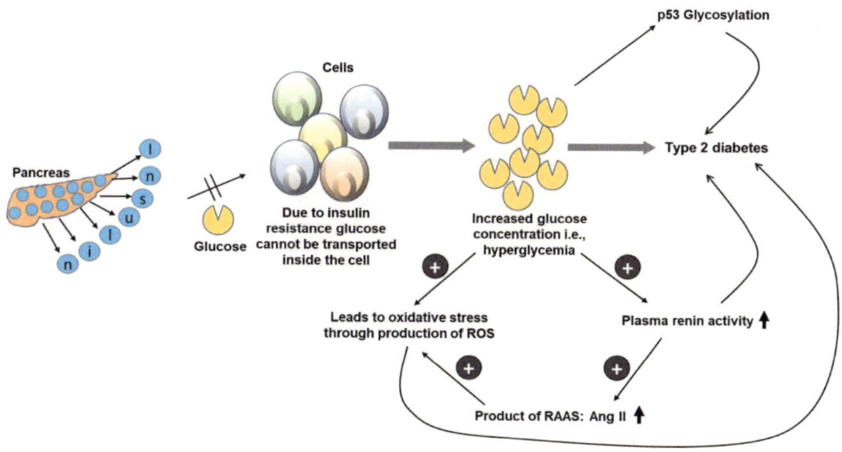

Figure 2.
Involvement of RAAS components into pathogenesis of T2DM. Hyperglycemia causes oxidative stress through generation of reactive oxygen species (ROS) that along with the production of Ang-II through overactive RAAS, may contribute to the pathophysiology of T2DM. Thus, genetic polymorphisms present in the genes expressing the components of RAAS probably modulate gene expression followed by protein levels that ultimately involve in the disease pathogenesis. Also, variants within these genes may also involve in the initiation and development of diabetes.

Another study revealed association between higher levels of aldosterone and insulin resistance along with dose-dependent contribution of high aldosterone level to the risk of developing T2D [51]. **Figure 2** schematically represents components of RAAS involved in the regulation of physiology, and probable mechanism of their contribution to the pathophysiology of diabetes.

5. T2DM and RAAS: contribution of the RAAS components to the pathogenesis of T2D

The most important key features of the pathogenesis of diabetes are the resistance to insulin which in turn reduces the insulin ability to uptake peripheral glucose [52], and the failure of β-cells to produce adequate amount of insulin [53]. Obesity is one of the major risk factors for the development of insulin resistance along with sedentary lifestyle, lack of physical activities etc. that in turn increases the levels of glucose in blood [54]. Obesity is also involved in the activation of RAAS [55, 56]. On the other hand, RAAS has been found to be associated with multiple obesity-associated chronic diseases, especially for cardiovascular related disease [57, 58]. In addition, several lines of evidence revealed association between activation of RAAS and the onset of T2D [55, 59, 60]. The connection between renin angiotensin system and insulin signaling pathway along with insulin resistance has been established [61]. A meta-analysis demonstrated that use of AT1R blockers or ACE inhibitors reduces the chance of new onset of T2DM by 22% in a population who are vulnerable to diabetes [62]. Though association between *ACE* I/D polymorphisms and risk of T2D inconsistent even in the same population [63, 64], CAPP trial demonstrated that ACE inhibitor captopril-treated patient group had 11% reduced chance of developing diabetes compared to non-treated groups [65] while LIFE study showed 25% reduction in new onset of diabetes [66]. All together these studies strongly support linkage between RAS components and hyperglycemia. Moderate hyperglycemia at the early stage of diabetes results increased plasma renin activity, arterial pressure and renal vascular resistance with the activation of both local and circulating RAAS [67, 68]. Moreover, hyperglycemia causes glycosylation of p53 which leads to the AGT transcription followed by the production of Ang-II [69, 70]. This was further supported by Fiordaliso et al. who demonstrated a direct correlation among levels of glucose, p53 glycosylation and Ang-II production [71].

Genetic predisposition involving certain SNPs residing within the genes of RAAS has been anticipated as the risk factors for the development and progression of T2D and T2D associated complications hypertension [72], coronary heart disease [73], nephropathy [73–75] and retinopathy [76]. Human AGT, a member of serpin gene family, comprises of 5 exons accounting for a full-length of about 12 kb and is situated on chromosome 1 (1q42-q43). Most convincing evidence for the probable association of polymorphic sites within *AGT* gene with essential hypertension has been identified in the 5′ flanking region, exons, and introns of the gene [77]. Strong association of rs11568020 (A-152G) and rs5050 (A-20C) in the promoter region as well as rs4762 and rs699 within exon 2 of *AGT* gene with hypertension was evident in Eastern Indian population [72]. Interestingly, incompatible findings with respect to the association of *AGT* variants with T2D have been observed [62, 72, 78, 79]. Variants rs699 and rs4762 within *AGT* gene found to be associated with the reduced risk of T2D in Eastern Indian and Malaysian Malays populations [72, 78] while no significant association was observed in the Chinese and the Japanese populations [63, 79]. However, rs699, rs4762 and rs5051 of *AGT* gene were reported to be associated with the increased risk of T2DM in the Pakistani [80], Korean [81] and Malaysian Malays [78] populations, respectively.

It has been documented that Ang-II is capable of stimulating the production of TGF-β [82] or inducing generation of reactive oxygen species (ROS) [83] that causes over-accumulation of extracellular matrix proteins or various cellular dysfunctions in patients with diabetes. Furthermore, variants present within the genes of RAAS components especially within *ACE*, *AGT* and *AT1R* genes have shown to be the most promising candidate genes susceptible to diabetic associated complications like nephropathy along with its progression towards renal failure as well as retinopathy [78, 84]. Haplotype TCG of *AGT* has been observed to be associated with increased risk of T2D [78]. According to Purkait et al. [72], three haplotypes (H4, H7 and H8) of *AGT* showed strong association with hypertension while H2 had protective role against this disease. It is reported that the *AT1R* A1166C is not likely a risk factor for chronic kidney disease in East Asians and Caucasians while it is shown to be a risk factor in South Asian population [85, 86]. Almost 30–50% of the diabetic individuals are prone to develop kidney disease [87, 88]. Previous studies reported association of *renin* gene polymorphisms with number of non-communicable diseases including diabetic nephropathy [89], increased risk of vascular complications [90], plasma renin activity [91], susceptibility to hypertension in a variety of ethnic groups [92–95], T2D [96] with inconsistent results [97–99]. Few studies did not find any significant association of *renin* rs16853055 with diabetes and diabetic nephropathy diseases [100, 101] while Purkait et al. [102] found an association of this variant with diabetic nephropathy in Indian population along with strong linkage disequilibrium with rs16853055. On the other hand, Deinum et al. reported weak association of *renin* gene polymorphism present in the first intron (involved in the regulation of transcription of renin) with diabetic nephropathy [89, 103]. Moreover, rs1799998 of the *CYP11B2* gene (aldosterone synthase) was associated with the levels of serum aldosterone and production [104, 105], blood pressure [106, 107], ischemic stroke [108], with the progression of renal function [109, 110] and end stage renal disease [111]. Meta-analysis performed by Xu et al. demonstrated association of allelic frequency as well as co-dominant homozygous and recessive models of inheritance with regard to −344 T/C polymorphism within promoter region of *CYP11B2* gene with the increased risk of diabetic nephropathy [112]. Similar association was observed by Purkait et al. [113] in Indian population. Promoter regions play important regulatory roles in gene transcription followed by formation of a functional protein through translation. Thus, presence of variants within the promoter region may be involved in the disease progression or pathogenicity which is definitely subject to further investigation and validation. Furthermore, methylation within the promoter region of a gene contributes to the expression of that particular gene [114]. Variant rs1799998 causes substitution of cytosine to thymidine within the promoter region of *CYP11B2* gene which is the binding site of a putative steroidogenic transcription factor-1 [115].

6. Pathophysiology and genetics of type 2 diabetes

Both environmental and genetic factors play pivotal role in the development of diabetes in human. However, some individuals develop diabetes while others do not although they use to live in the same environment. A substantial proportion of Pima Indians develop T2D even with a normal lifestyle in a normal environment that showed strong linkage of genetic make-up to T2D [116]. Thus, understanding genetics related to the pathogenesis of T2D is of utmost importance for the management of this global endemic disease. Familial studies orchestrated more robust data as proof that genes play important role as risk factor for the development of diabetes. First degree individuals with family history of T2D are at 3-fold increased risk of developing T2D compared to those who do not have positive family history [117–119]. Studies with monozygotic twins demonstrated that 50% risk of developing type 1 diabetes is contributed by *HLA* genes while rest of the 50% is associated with environmental factors and epigenetic modifications [120, 121]. Several family, population and twin-based studies established that heritability of T2D ranges from 20–80% [122, 123]. Forty percent individuals possess risk of developing T2D who have one parent with T2D while 70% of the individuals have higher risk of developing T2D if both the parents are T2D [124]. Seventy percent of monozygotic twins are in concordance with the chance of developing T2D while the concordance rate in dizygotic twins has been found to be 20–30% [125, 126].

The primary method to identify genes susceptible to T2D was genome linkage analysis. This approach efficiently identified causal mutations specially for the monogenic forms of diabetes like maturity-onset diabetes in young (MODY), mitochondrial diabetes in neonates and insulin resistance [127–129]. This approach further recognized the short tandem repeats located on q arm of chromosomes 4, 5, 10, 12, 22 and p arm of chromosomes 2, 3, 6, 13 for their probable association with T2D in different ethnic populations [130–134] along with causative genetic variants within *calpain10* (*CAPN10*) [135], *ENPP1* [136], *HNF4A* [137, 138] and *ACDC* [139]. Calcium-activated neutral protease 10, one of the regulator of glucose homeostasis, gene (*CAPN10*) variants UCSNP-43 G/A in intron 3, UCSNP-19 2R (two 32-bp repeats)/3R (three 32-bp repeats) in intron 6 and UCSNP-63 C/T in intron 13 have been reported to be associated with T2D in Mexicans Americans, German and Finnish populations [135, 140]. The ectonucleotide pyrophosphatase phosphodiesterase (ENPP1) was supposed to be associated with insulin resistance [141]. The three-alleles risk haplotype (K121Q/IVS20 delT-11/A > G + 1044 TGA, QdelTG) within ENPP1 was associated with childhood obesity, development of T2D and with adult obesity [136]. HNF4A, member of the steroid hormone receptor superfamily, plays major role in insulin expression and secretion followed by glucose metabolism in pancreatic β-cells along with gluconeogenesis in liver [142, 143]. Variants within *HNF4A* gene were identified as the risk factor for MODY and causative factor for β-cell dysfunctions [144]. Also, non-coding variants rs4812829 and rs6017317 as well as coding variant rs1800961 (T130I) within *HNF4A* were involved in the development of T2D [145–147]. Decreased level of adipose tissue-derived adiponectin in plasma is evident in individuals with obesity [148], insulin resistance [149] and T2D [148]. Adiponectin encoding *ACDC* gene variants 276G > T and 45 T > G were found to be associated with lower levels of plasma adiponectin in Japanese [150] and German obese people [151], respectively along with their predisposition to T2D. However, genome wide linkage analyses did not reveal any association of these variants of *ACDC* gene with obesity and T2D in Pima Indians [139]. Transcription factor TCF7L2 showed strongest linkage to the risk of T2D before genome wide association study (GWAS) era [130]. TCF7L2 involves in Wnt signaling pathway that regulates proliferation and survival of pancreatic islet cell functions [152] and its reduced expression is linked to impaired insulin secretion [153]. *TCF7L2* gene variants rs12255372 and rs7903146, showed strong linkage disequilibrium with composite at-risk alleles of the microsatellite marker (DG10S478).

Candidate gene association studies have also been proved to be effective to obtain substantial evidences of genetic predisposition to T2D. For example, insulin-like growth factor 2 mRNA-binding protein 2 (IGF2BP2), an important candidate gene for T2D [154, 155], was involved with T2D development by reducing insulin secretion [156] may be through changing adipose tissue and β-cell function [157]. IGF2BP2 was also associated with overweight and obesity [158]. Association of rs4402960 and rs1470579 within *IGF2BP2* with the risk of T2D demonstrated in French Caucasians while another study revealed that T2D patients carrying the

T allele of rs4402960 had higher levels of fasting plasma glucose, postprandial glucose, total cholesterol and postprandial serum insulin compared to individuals with the GG genotype [158]. Besides, *IGF2BP2* variants showed effect on treatment of diabetes. For example, lower efficacy of the repaglinide treatment for reducing fasting plasma glucose and postprandial glucose was observed in diabetic patients with rs1470579 AC + CC genotypes compared to AA genotypes. On the other hand, repaglinide treatment had higher effect on diabetic patients with GT + TT genotypes with regard to rs4402960 on postprandial insulin compared to GG genotype carrying patients [158]. The potassium inwardly rectifying channel, subfamily J, member 11 (KCNJ11) has attracted attention due to its contribution to the pathogenesis of T2D by modulating insulin production and secretion [159] and thus, is a good candidate gene to elucidate its disease association. *KCNJ11* harboring four missense SNPs rs5219, rs1800467, rs5215, rs41282930 were recognized to influence risk of T2D by impairing insulin secretion [160]. Peroxisome proliferator activator receptor gamma (PPARG) was identified to harbor T2D disease susceptibility variants. Both *KCNJ11* and *PPARG* encode anti-diabetic drug targets and their respective missense SNPs rs5219 (E23K) and rs1801282 (P12A) are associated with the risk of T2D [161].

Although candidate gene and linkage analyses provided considerable evidences behind the genes for their probable association with the pathophysiology of T2D and/or with the risk of T2D, novel genes are yet demanding due to the inconsistent and discordant findings within the same population and also, in different ethnic groups. Screening of whole genome using GWAS helps to overcome the shortcomings of the above mentioned approaches to some extent by expediting regularly spaced variants without any prior knowledge of gene or their effects that has brought a significant breakthrough in understanding the genetic basis of T2D. This has become realistic after successful completion of the Human Genome Project and the International HapMap Project. This has given an opportunity to deposit millions of SNPs in the public databases [162] and presence of higher frequency of a particular SNP in cases compare to controls suggests association of that SNP with the case i.e., disease. Moreover, to satisfy association of SNPs statistically, stringent p value ($<10^{-8}$) is required in GWAS and it benefited researchers to eliminate false positive association out of the millions of reported SNPs [163]. Even with such strict threshold levels of statistics, several case–control studies in different ethnicities have generated replicative positive results through different independent datasets. T2D associated variants within genes uncovered by GWAS positioned at different chromosomal locations (**Figure 3A**) can be grouped into i) insulin secretion and processing related (*GIPR*, *CCND2*, *CDKAL1*, *GCK*, *TCF7L2*, *GLIS3*, *THADA*, *IGF2BP2*, *DGKB*), ii) impaired insulin function related *(PPARG, KLF14, IRS1),* iii) insulin resistance related (*ACDC, FTO, KLF14, DUSP9*), iv) β-cell mass and function related (*IGF2BP2, HCNQ1, CDKN2A, CDKN2B*) and iii) body mass index (BMI) and lipid level related (*NRXN3, CMIP, APOE, and* MC4R). Notably rs4731702 of intronless *KLF14* demonstrated an association with insulin resistance [164] while rs972283 contributed to elevated blood pressure [165], which may ultimately increase risk of cardiovascular disease; C allele of the rs2283228 within *HCNQ1* showed association with increased fasting glucose levels and impaired β-cell function in Asians [166], while C allele of rs2237895 in *KCNQ1* was found to be related to decreased risk of abdominal obesity in patients with T2DM [167, 168]; rs5945326 of *DUSP9* on X chromosome was related to the increased risk of T2D in Japanese [169], Pakistanis [170] and in European [171] populations; rs1558902 within *FTO* showed correlation with the incidence T2D in humans even after adjusting the data with confounding factors such as age and BMI [172] and rs9939609 may modulate the risk of T2D by regulating other genes, an incidence

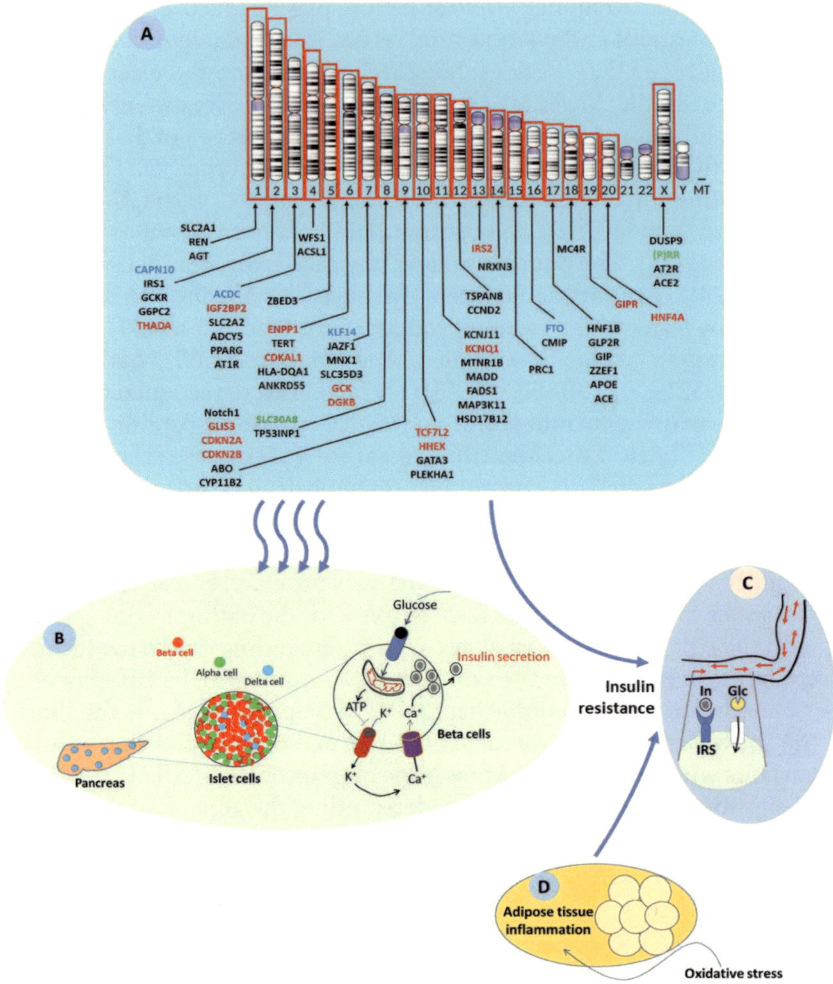

Figure 3.

Chromosomal locations of genes carrying variants (A) associated with β-cell function followed by insulin production and secretion (B), glucose utilization and homeostasis (C) along with glycemic traits and abnormal adipose tissue function (D) which all together may lead to T2D and of genes of major RAAS components. Several approaches specially GWASs identified several variants associated with pancreatic islet cell function followed by β-cell dysfunction, insulin secretion and processing (red)*, with development of insulin resistance followed by imbalanced glucose homeostasis* (blue)*. Other variants are also associated with abnormal adipose tissue function which may also be caused by oxidative stress, a consequence of Ang-II (**Figure 2**). Variants within SLC30A8 and (P)RR* (green) *showed both protection against T2D and risk association with T2D as well as hypertension, respectively. Also, mostly non-coding and few coding variants within the genes* (black) *showed association with the risk of T2D. Variants within the major gene of RAAS have been found to be associated with the risk of T2D and T2D-associated hypertension other that their established risk association with essential hypertension and cardiovascular diseases. REN, renin; AGT, angiotensinogen; AT1R and AT2R, angiotensin type 1 and type 2 receptors, ACE, angiotensin converting enzyme; ACE2, angiotensin converting enzyme 2; CYP11B2, aldosterone synthase; (P)RR, (pro)renin receptor; In, insulin; Glc, glucose; IRS, insulin receptor substrate.*

independent of BMI [173]; variants present within the tumor suppressor cyclin dependent kinase inhibitors, CDKN2A and CDKN2B, reported to be associated with T2D in Asians and Europeans [174–177]. rs10811661 of CDKN2A/2B is also, according to GWAS, linked to diabetes [178]; hematopoietically-expressed homeobox or *HHEX* gene variants rs11118745G/A, rs7923837A/G, and rs5015480C/T had been identified as risk factors for T2D in Japanese [179], German [180], Korean [181], Indian [182] populations. Association of a common variant, Trp325Arg within *SLC30A8*, with the risk of T2D [171, 183] and, levels of glucose [184] and proinsulin [185] had been well documented. Interestingly, through genotyping of ~150,000 individuals from five ethic groups, Flannick et al. (2014) revealed protective role against the development of T2D mediated by the loss of function variants harbored within *SLC30A8* [186]. AA genotype of rs11558471 of *SLC30A8* was found significantly more frequent in T2D patients than in controls in Han Chinese [187] and Indian [182] populations.

Non-coding variants within different genes [like variants of *PRC1,* MADD, *MTNR1B,* FADS1, CRY2, *GLIS3, LC2A2, ADCY5,* GCKR, G6CP2 [184], *TP53INP1* [188], *GIPR* [189], *ADCY5* [189], *TSPAN8/LGR5, JAZF1,* Notch1 [190], *HNF1B* [191], *FTO* [155], *ZEDB3* [188]], as presented in **Figure 3A**, were also recognized as major risk factors associated with the development of T2D and/or regulation of glucose/insulin homeostasis, and/or glycemic traits (**Figure 3B** and **C**) and abnormal adipose tissue function (**Figure 3D**) while few variants were discerned to have protective roles against the development of diabetes [184, 188–195]. Also, similar association was found with regard to intergenic variants rs972283 (G/A, 47 kb upstream) of *KLF14* [188], rs2943641 (C/T, 502 kb upstream) of *IRS-1* [155], rs1111875 (C/T, 7.7 kb downstream) of *HHEX* [183], rs10811661 (T/C, 125 kb upstream) of *CDKN2A/2B [190],* rs4607103 (C/T, 38 kb upstream) of *ADAMSTS9* [190], *regulatory region variant* rs5945326 (G/A, 8 kb upstream) of *DUSP9* [188], rs2191349 (T/G) of *DGKB/TMEM195* [196], promoter region rs2853669 of human telomerase reverse transcriptase (*TERT*) gene [197]. Non-coding variants positioned at essential regions like enhancer and promoter sequence may also modulate chromatin loops, alter sequence motifs and modulate histone marks that ultimately regulate gene expression, which could be one of the key reasons their disease association.

7. Pathogenesis and genetics of RAAS

RAAS is the enzymatic cascade to produce the effector molecule, Ang-II, by the multiple enzymes [23] (**Figure 1**). Various genotypes of the RAAS components [eg. AGT, renin, ACE, ACE2, AT1R, AT2R and (P)RR] have been investigated to find the link between genetic variation, blood pressure, and hypertension [198].

The two *AGT* genotypes (G-6A non-coding SNP and M235T coding SNP) are associated with higher plasma AGT levels and increased risks of essential hypertension [77]. The *AGT* SNPs occurring within the non-coding region could explain the association with plasma AGT concentration because of the alternation in AGT transcription [198]. It is plausible that the higher AGT concentration brings about the higher levels of Ang-II, which may lead to high blood pressure. In the study of 10,690 individuals, the associations of elevated blood pressure, ischemic heart disease and ischemic cerebrovascular disease were examined with four AGT variants (A-20C and G-6A non-coding SNPs and T174M and M235T coding SNPs) [199]. Both women and men with -6AA, 174TT, and 235TT (versus -6GG, 174TT, and 235TT) had higher mean levels of plasma AGT (861 ng/mL and 811 ng/mL, respectively). This finding suggests that the genotype has an effect on risk of elevated blood pressure in women, but not in men [199]. The association of the genotype with ischemic heart disease and ischemic cerebrovascular disease seems weak as a risk [199]. A meta-analysis of 45,267 individuals from different ethnic populations shows that M235T genotype is associated with an increase in plasma AGT levels [200]. An analysis of 424 individuals from 41 two-generation families from Utah indicates significant linkage between six *AGT* SNPs (rs5051, rs699, rs6687360, rs2478543, rs3789670 and rs943580) and plasma AGT levels whereas plasma AGT and blood pressure were not significantly correlated [201]. *AGT* SNPs have been identified from various ethnic groups to show its association with hypertension [72, 202–205]. Of note, *AGT* genotypes (G-6A, T + 31C and M235T) with hypertension are not associated with plasma AGT level, while -1074 t|T235 haplotype is associated with an increase of AGT level but not with hypertension [202]. Sato et al. [202] suggested that the positive association between *AGT* polymorphism and hypertension is not simply explained by an increase of plasma AGT concentration.

Renin polymorphism was investigated by assessing the association of ten *renin* genotypes with hypertension risk in 570 hypertensive and 222 normotensive Caucasians [95]. Subjects with DM, secondary hypertension, significant medical illness or severe obesity were excluded, and their food intakes were also controlled. The A allele of rs6693954 SNP and the haplotype containing rs6693954A were significantly associated with higher risk of hypertension [95]. Compared to other haplotypes, the same haplotype showed the higher levels of plasma renin activity, suggesting that a direct renin inhibitor is effective to reduce blood pressure of rs6693954A carriers [95]. In addition, the haplotype displayed a blunted mean arterial pressure response to exogenously infused Ang-II [95], which infers the dysregulation of RAAS at the tissue level [206]. This study [95] confirms the association between *renin* genotypes and risk for hypertension.

As described above, genetic variations in individual RAAS components can contribute to the onset of physiological outcomes, which probably brings about the increase in blood pressure. But hypertension is a multifactorial disease involving both genetic and environmental factors [207] like T2D. The mechanism of susceptibility to hypertension and CVD is much more complex, since various genes work in an additive or interactive manner, together with environmental factors [198]. Ji et al. [205] provided the experimental evidence to support the idea. In a study of 905 hypertensive and 905 normotensive Han Chinese population, 41 SNPs of the five RAAS components (AGT, renin, ACE, AT1R, and CYP11B2) and the non-genetic factors were analyzed to investigate their associations with essential hypertension [205]. Subjects with CVD, DM, kidney diseases, secondary hypertension and other major chronic illnesses were excluded. Serum levels of total cholesterol and triglyceride, and BMI were significantly higher in the hypertensive group than in the normotensive group. Six SNPs (rs3789678 and rs2493132 within *AGT*, rs4305 within *ACE*, rs275645 within *AT1R*, rs3802230 and rs10086846 within *CYP11B2*) were shown to associate with hypertension. The interaction between BMI and rs4305 (*ACE* SNPs) increased the susceptibility to hypertension. Together with non-genetic factors, the genetic variations in the RAAS components may play an important role in determining an individual's susceptibility to hypertension [205].

GWAS analysis performed by Ji et al. [208] provided one important viewpoint on genetic polymorphism of RAAS. The authors searched GWAS Catalog (https://www.ebi.ac.uk/gwas/) and identified all known RAAS genes and relevant diseases and traits. Remarkably, SNPs within *AGT*, *renin*, *ACE2*, *CYP11B2*, *ATP6AP2* [*(P)RR*] and *HSD11B2* were not associated with any disease and trait. There were SNPs being associated with other disease and trait: *ACE* (metabolic traits), *AT1R* (leads levels in blood), *AT2R* (fibrosis), *MAS1* (lipoprotein levels), *RENBP* (schizophrenia) and *NR3C2* (thyroid function). But these six SNPs showed no direct association with hypertension. The only SNP associated with a blood pressure trait was rs17367504, which is located in the intronic region of methylenetetrahydrofolate reductase (*MTHFR*) gene near many plausible candidate genes, including ion channel *CLCN6*, natriuretic peptides *NPPA* and *NPPB*, and RAAS component *AGTRAP*. The authored emphasized that the contribution of RAAS variants needs to be reconsidered when evaluating one's susceptibility of hypertension [208]. GWAS analysis is providing a new dimension for understanding genetic architecture of blood pressure and Page's "mosaic theory" of hypertension [209].

SARS-CoV-2 has emerged in December 2019, which caused COVID-19. The SARS-CoV-2 spike protein directly binds to ACE2, which is present on lung epithelial cells and other tissues [210]. ACE2 converts Ang-II to Ang 1–7 leading to tissue repair signal (**Figure 1**). When SARS-CoV-2 is attached to ACE2, it likely reduces the ACE2 activity associated with reduced inflammation, thereby increasing lung injury due to the decrease in Ang 1–7 generation [210]. It was observed that the severe COVID-19 patients are likely to have a history of diabetes, hypertension or CVD [5, 6]. For reducing the infection by COVID-19 and the other coronaviruses, deciphering the susceptibility to hypertension in term of genetic variations should be indispensable, which will be achieved by steady efforts to clarify the genetic background of each ethnic.

We recently reported probable association of five non-coding SNPs within *renin* and *(P)RR* genes with T2D, hypertension and T2D-associated hypertension in Bangladeshi population [211]. *Renin* SNP rs3730102 was associated with an increased risk of the three diseases. *Renin* SNP rs11571079 was associated with an increased risk for hypertension and T2D-associated hypertension, while the SNP showed a decreased risk for T2D, exerting a protective effect. *(P)RR* rs2968915|rs3112298 haplotypes were related to an increased risk of T2D and T2D-associated hypertension. These findings highlight important roles of non-coding variants of *renin* and *(P)RR* genes in the etiology of several polygenic diseases [211]. Although there is a limitation for genotyping the candidate SNPs for the disease risk prediction, finding the candidate gene in different ethnic group through "one-to-one" approach should be valuable to design a measure for ensuring health and quality of life at all ages in each population group.

8. Conclusion

Though several studies have revealed genetic approaches to identify the pathophysiology of diabetes, hypertension and/or diabetes associated complications, it is still very challenging to uncover a definite candidate for the genetic etiology of these diseases due to overlapping involvement of genes, loci or even SNPs. GWASs have come forward to get rid of this elusiveness through scanning of whole genome. However, it is still very challenging due to the ethnic variations and ethnicity-dependent gene expression patterns even harboring the same loci and/or variants to recognize genetic risk factors. Rather panels of variants (panels of variants for more closely related to T2D, panels for more closely related to hypertension and panels of overlapping variants in case of T2D and hypertension) could be a more meticulously related suggestive diagnostic, predictive and prognostic biomarker for these diseases. Known variants along with their gene expression pattern may play a pivotal role in determining disease pathogenesis.

Acknowledgements

We are grateful to all the members of Laboratory of Population Genetics, University of Dhaka as well as Biological Chemistry Laboratory and Biomolecular Chemistry Laboratory, Gifu University for their valuable suggestions and support. This work was supported in part by the JSPS KAKENHI (Grant No. 15 K01707 and 18KK0273).

Conflict of interest

The authors declare no competing interests.

Abbreviations

Ang 1–7	angiotensin 1–7;
ACE	angiotensin converting enzyme;
ACE2	angiotensin converting enzyme 2;
AGT	angiotensinogen;
Ang-I	angiotensin-I;
Ang-II	angiotensin-II;
AT1R	angiotensin type 1 receptor;
AT2R	angiotensin type 2 receptor;
COVID-19	coronavirus infectious disease 2019;
CVD	cardiovascular disease;
DM	diabetes mellitus;
ENPP1	ectonucleotide pyrophosphatase phosphodiesterase;
GPCR	G-protein coupled receptor;
GWAS	genome wide association study;
I/D	insertion/deletion;
IGF2BP2	insulin-like growth factor 2 mRNA-binding protein 2;
IRS-1	insulin receptor substrate-1;
KCNJ11	potassium inwardly rectifying channel subfamily J member 11;
MAS1	MAS-related;
MODY	maturity-onset diabetes in young;
(P)RR	(pro)renin receptor;
RAAS	renin-angiotensin-aldosterone system;
ROS	reactive oxygen species;
SARS-CoV-2	severe acute respiratory syndrome coronavirus 2;
SNP	single nucleotide polymorphism;
T2D	type 2 diabetes;
T1DM	type 1 diabetes mellitus;
T2DM	type 2 diabetes mellitus;
PPARG	peroxisome proliferator activator receptor gamma;
TERT	telomerase reverse transcriptase

Author details

A.H.M. Nurun Nabi[1,2] and Akio Ebihara[3,4,5]*

1 Laboratory of Population Genetics, Department of Biochemistry and Molecular Biology, University of Dhaka, Dhaka, Bangladesh

2 United Graduate School of Agricultural Sciences, Gifu University, Gifu, Japan

3 Faculty of Applied Biological Sciences, Gifu University, Gifu, Japan

4 Center for Highly Advanced Integration of Nano and Life Sciences, Gifu University (G-CHAIN), Gifu, Japan

5 Department of Chemical Engineering, Indian Institute of Technology Guwahati, Guwahati, Assam, India

*Address all correspondence to: aebihara@gifu-u.ac.jp

References

[1] Zimmet PZ, Magliano DJ, Herman WH, Shaw JE. Diabetes: a 21st century challenge. Lancet Diabetes Endocrinol. 2014;2:56-64. DOI:10.1016/S2213-8587(13)70112-8

[2] Yang MM, Wang J, Ren H, Sun YD, Fan JJ, Teng Y, et al. Genetic Investigation of Complement Pathway Genes in Type 2 Diabetic Retinopathy: An Inflammatory Perspective. Mediators Inflamm. 2016;2016:1313027. DOI:10.1155/2016/1313027

[3] Oparil S, Acelajado MC, Bakris GL, Berlowitz DR, Cifkova R, Dominiczak AF, et al. Hypertension. Nat Rev Dis Primers. 2018;4:18014. DOI:10.1038/nrdp.2018.14

[4] Mendis S, Puska P, Norrving B. Global atlas on cardiovascular disease prevention and control. Geneva, Switzerland: World Health Organization; 2011. DOI: https://doi.org/10.1016/S0022-3476(51)80275-0

[5] Singh AK, Gupta R, Ghosh A, Misra A. Diabetes in COVID-19: Prevalence, pathophysiology, prognosis and practical considerations. Diabetes Metab Syndr. 2020;14:303-310. DOI:10.1016/j.dsx.2020.04.004

[6] de Almeida-Pititto B, Dualib PM, Zajdenverg L, Dantas JR, de Souza FD, Rodacki M, et al. Severity and mortality of COVID 19 in patients with diabetes, hypertension and cardiovascular disease: a meta-analysis. Diabetol Metab Syndr. 2020;12:75. DOI:10.1186/s13098-020-00586-4

[7] Federation ID. IDF Diabetes Atlas, 9th edn. Brussels, Belgium; 2019.

[8] Guariguata L, Whiting DR, Hambleton I, Beagley J, Linnenkamp U, Shaw JE. Global estimates of diabetes prevalence for 2013 and projections for 2035. Diabetes Res Clin Pract. 2014;103:137-149. DOI:10.1016/j.diabres.2013.11.002

[9] Ramachandran A, Snehalatha C, Ma RC. Diabetes in South-East Asia: an update. Diabetes Res Clin Pract. 2014;103:231-237. DOI:10.1016/j.diabres.2013.11.011

[10] American Diabetes Association. 2. Classification and Diagnosis of Diabetes: Standards of Medical Care in Diabetes-2021. Diabetes Care. 2021;44:S15-S33. DOI:10.2337/dc21-S002

[11] Morgan CL, Currie CJ, Peters JR. Relationship between diabetes and mortality: a population study using record linkage. Diabetes Care. 2000;23:1103-1107. DOI:10.2337/diacare.23.8.1103

[12] Haffner SM, Lehto S, Ronnemaa T, Pyorala K, Laakso M. Mortality from coronary heart disease in subjects with type 2 diabetes and in nondiabetic subjects with and without prior myocardial infarction. N Engl J Med. 1998;339:229-234. DOI:10.1056/NEJM199807233390404

[13] Goodkin DA, Bragg-Gresham JL, Koenig KG, Wolfe RA, Akiba T, Andreucci VE, et al. Association of comorbid conditions and mortality in hemodialysis patients in Europe, Japan, and the United States: the Dialysis Outcomes and Practice Patterns Study (DOPPS). J Am Soc Nephrol. 2003;14:3270-3277. DOI:10.1097/01.asn.0000100127.54107.57

[14] Rao Kondapally Seshasai S, Kaptoge S, Thompson A, Di Angelantonio E, Gao P, Sarwar N, et al. Diabetes mellitus, fasting glucose, and risk of cause-specific death. N Engl J Med. 2011;364:829-841. DOI:10.1056/NEJMoa1008862

[15] Casqueiro J, Casqueiro J, Alves C. Infections in patients with diabetes mellitus: A review of pathogenesis. Indian J Endocrinol Metab. 2012;16 Suppl 1:S27–S36. DOI:10.4103/2230-8210.94253

[16] Geerlings SE, Hoepelman AI. Immune dysfunction in patients with diabetes mellitus (DM). FEMS Immunol Med Microbiol. 1999;26:259-265. DOI:10.1111/j.1574-695X.1999.tb01397.x

[17] Islam LN, Islam KS, Kabir ME, Hossain MS, Nabi A, Banik NG. Impaired neutrophil function in diabetes mellitus. Bangladesh J Biochem. 2000;6:45-59

[18] Peleg AY, Weerarathna T, McCarthy JS, Davis TM. Common infections in diabetes: pathogenesis, management and relationship to glycaemic control. Diabetes Metab Res Rev. 2007;23:3-13. DOI:10.1002/dmrr.682

[19] Nabi AH, Islam LN, Rahman MM, Biswas KB. Polymorphonuclear neutrophil dysfunctions in streptozotocin-induced type 1 diabetic rats. J Biochem Mol Biol. 2005;38:661-667. DOI:10.5483/bmbrep.2005.38.6.661

[20] Muller LM, Gorter KJ, Hak E, Goudzwaard WL, Schellevis FG, Hoepelman AI, et al. Increased risk of common infections in patients with type 1 and type 2 diabetes mellitus. Clin Infect Dis. 2005;41:281-288. DOI:10.1086/431587

[21] Deng SQ, Peng HJ. Characteristics of and Public Health Responses to the Coronavirus Disease 2019 Outbreak in China. J Clin Med. 2020;9. DOI:10.3390/jcm9020575

[22] Dooley KE, Chaisson RE. Tuberculosis and diabetes mellitus: convergence of two epidemics. Lancet Infect Dis. 2009;9:737-746. DOI:10.1016/S1473-3099(09)70282-8

[23] Nabi AN, Biswas KB, Ebihara A, Nakagawa T, Suzuki F. Renin angiotensin system in the context of renin, prorenin, and the (pro)renin receptor. Reviews in Agricultural Science. 2013;1:43-60. DOI:10.7831/ras.1.43

[24] Te Riet L, van Esch JH, Roks AJ, van den Meiracker AH, Danser AH. Hypertension: renin-angiotensin-aldosterone system alterations. Circ Res. 2015;116:960-975. DOI:10.1161/CIRCRESAHA.116.303587

[25] Nabi AN, Suzuki F, Inagami T. Renin. In: Rawlings ND, Salvesen GS, editors. Handbook of Proteolytic Enzymes. 1. 3rd ed: Academic Press, Oxford; 2013. p. 71-82.

[26] Riordan JF. Angiotensin-I-converting enzyme and its relatives. Genome Biol. 2003;4:225. DOI:10.1186/gb-2003-4-8-225

[27] Ito M, Oliverio MI, Mannon PJ, Best CF, Maeda N, Smithies O, et al. Regulation of blood pressure by the type 1A angiotensin II receptor gene. Proc Natl Acad Sci U S A. 1995;92:3521-3525. DOI:10.1073/pnas.92.8.3521

[28] Siragy HM, Carey RM. The subtype 2 (AT2) angiotensin receptor mediates renal production of nitric oxide in conscious rats. J Clin Invest. 1997;100:264-269. DOI:10.1172/JCI119531

[29] Donoghue M, Hsieh F, Baronas E, Godbout K, Gosselin M, Stagliano N, et al. A novel angiotensin-converting enzyme-related carboxypeptidase (ACE2) converts angiotensin I to angiotensin 1-9. Circ Res. 2000;87:E1–E9. DOI:10.1161/01.res.87.5.e1

[30] Karnik SS, Singh KD, Tirupula K, Unal H. Significance of angiotensin 1-7 coupling with MAS1 receptor and other GPCRs to the renin-angiotensin system: IUPHAR Review 22. Br J Pharmacol.

2017;174:737-753. DOI:10.1111/bph.13742

[31] Leung KK, Leung PS. Effects of hyperglycemia on angiotensin II receptor type 1 expression and insulin secretion in an INS-1E pancreatic beta-cell line. JOP. 2008;9:290-299

[32] Sowers JR. Insulin resistance and hypertension. Am J Physiol Heart Circ Physiol. 2004;286:H1597–H1602. DOI:10.1152/ajpheart.00026.2004

[33] Luther JM, Brown NJ. The renin-angiotensin-aldosterone system and glucose homeostasis. Trends Pharmacol Sci. 2011;32:734-739. DOI:10.1016/j.tips.2011.07.006

[34] Underwood PC, Adler GK. The renin angiotensin aldosterone system and insulin resistance in humans. Curr Hypertens Rep. 2013;15:59-70. DOI:10.1007/s11906-012-0323-2

[35] Hitomi H, Kiyomoto H, Nishiyama A, Hara T, Moriwaki K, Kaifu K, et al. Aldosterone suppresses insulin signaling via the downregulation of insulin receptor substrate-1 in vascular smooth muscle cells. Hypertension. 2007;50:750-755. DOI:10.1161/HYPERTENSIONAHA.107.093955

[36] Krug AW, Ehrhart-Bornstein M. Aldosterone and metabolic syndrome: is increased aldosterone in metabolic syndrome patients an additional risk factor? Hypertension. 2008;51:1252-1258. DOI:10.1161/HYPERTENSIONAHA.107.109439

[37] Nguyen G, Delarue F, Burckle C, Bouzhir L, Giller T, Sraer JD. Pivotal role of the renin/prorenin receptor in angiotensin II production and cellular responses to renin. J Clin Invest. 2002;109:1417-1427. DOI:10.1172/JCI14276

[38] Nabi AH, Kageshima A, Uddin MN, Nakagawa T, Park EY, Suzuki F. Binding properties of rat prorenin and renin to the recombinant rat renin/prorenin receptor prepared by a baculovirus expression system. Int J Mol Med. 2006;18:483-488

[39] Nabi AH, Biswas KB, Nakagawa T, Ichihara A, Inagami T, Suzuki F. Prorenin has high affinity multiple binding sites for (pro)renin receptor. Biochim Biophys Acta. 2009;1794:1838-1847. DOI:10.1016/j.bbapap.2009.08.024

[40] Nabi AH, Biswas KB, Nakagawa T, Ichihara A, Inagami T, Suzuki F. 'Decoy peptide' region (RIFLKRMPSI) of prorenin prosegment plays a crucial role in prorenin binding to the (pro)renin receptor. Int J Mol Med. 2009;24:83-89. DOI:10.3892/ijmm_00000210

[41] Hirose T, Hashimoto M, Totsune K, Metoki H, Asayama K, Kikuya M, et al. Association of (pro)renin receptor gene polymorphism with blood pressure in Japanese men: the Ohasama study. Am J Hypertens. 2009;22:294-299. DOI:10.1038/ajh.2008.357

[42] Hirose T, Hashimoto M, Totsune K, Metoki H, Hara A, Satoh M, et al. Association of (pro)renin receptor gene polymorphisms with lacunar infarction and left ventricular hypertrophy in Japanese women: the Ohasama study. Hypertens Res. 2011;34:530-535. DOI:10.1038/hr.2010.274

[43] Sakoda M, Ichihara A, Kaneshiro Y, Takemitsu T, Nakazato Y, Nabi AH, et al. (Pro)renin receptor-mediated activation of mitogen-activated protein kinases in human vascular smooth muscle cells. Hypertens Res. 2007;30:1139-1146. DOI:10.1291/hypres.30.1139

[44] Ramser J, Abidi FE, Burckle CA, Lenski C, Toriello H, Wen G, et al. A unique exonic splice enhancer mutation in a family with X-linked mental retardation and epilepsy points to a novel role of the renin receptor. Hum

Mol Genet. 2005;14:1019-1027. DOI:10.1093/hmg/ddi094

[45] George AJ, Thomas WG, Hannan RD. The renin-angiotensin system and cancer: old dog, new tricks. Nat Rev Cancer. 2010;10:745-759. DOI:10.1038/nrc2945

[46] Wegman-Ostrosky T, Soto-Reyes E, Vidal-Millan S, Sanchez-Corona J. The renin-angiotensin system meets the hallmarks of cancer. J Renin Angiotensin Aldosterone Syst. 2015;16:227-233. DOI:10.1177/1470320313496858

[47] Deckers IA, van den Brandt PA, van Engeland M, van Schooten FJ, Godschalk RW, Keszei AP, et al. Polymorphisms in genes of the renin-angiotensin-aldosterone system and renal cell cancer risk: interplay with hypertension and intakes of sodium, potassium and fluid. Int J Cancer. 2015;136:1104-1116. DOI:10.1002/ijc.29060

[48] Group UPDSU. Intensive blood-glucose control with sulphonylureas or insulin compared with conventional treatment and risk of complications in patients with type 2 diabetes (UKPDS 33). Lancet. 1998;352:837-853

[49] Holman RR, Paul SK, Bethel MA, Matthews DR, Neil HA. 10-year follow-up of intensive glucose control in type 2 diabetes. N Engl J Med. 2008;359:1577-1589. DOI:10.1056/NEJMoa0806470

[50] Reincke M, Meisinger C, Holle R, Quinkler M, Hahner S, Beuschlein F, et al. Is primary aldosteronism associated with diabetes mellitus? Results of the German Conn's Registry. Horm Metab Res. 2010;42:435-439. DOI:10.1055/s-0029-1246189

[51] Joseph JJ, Echouffo-Tcheugui JB, Kalyani RR, Yeh HC, Bertoni AG, Effoe VS, et al. Aldosterone, Renin, and Diabetes Mellitus in African Americans: The Jackson Heart Study. J Clin Endocrinol Metab. 2016;101:1770-1778. DOI:10.1210/jc.2016-1002

[52] Reaven GM. Banting lecture 1988. Role of insulin resistance in human disease. Diabetes. 1988;37:1595-1607. DOI:10.2337/diab.37.12.1595

[53] Porte D, Jr. Banting lecture 1990. Beta-cells in type II diabetes mellitus. Diabetes. 1991;40:166-180. DOI:10.2337/diab.40.2.166

[54] Adeghate E, Schattner P, Dunn E. An update on the etiology and epidemiology of diabetes mellitus. Ann N Y Acad Sci. 2006;1084:1-29. DOI:10.1196/annals.1372.029

[55] Goossens GH, Blaak EE, van Baak MA. Possible involvement of the adipose tissue renin-angiotensin system in the pathophysiology of obesity and obesity-related disorders. Obes Rev. 2003;4:43-55. DOI:10.1046/j.1467-789x.2003.00091.x

[56] Engeli S, Negrel R, Sharma AM. Physiology and pathophysiology of the adipose tissue renin-angiotensin system. Hypertension. 2000;35:1270-1277. DOI:10.1161/01.hyp.35.6.1270

[57] Unger T. The role of the renin-angiotensin system in the development of cardiovascular disease. Am J Cardiol. 2002;89:3A-9A; discussion 10A. DOI:10.1016/s0002-9149(01)02321-9

[58] Ferrario CM, Strawn WB. Role of the renin-angiotensin-aldosterone system and proinflammatory mediators in cardiovascular disease. Am J Cardiol. 2006;98:121-128. DOI:10.1016/j.amjcard.2006.01.059

[59] Olivares-Reyes JA, Arellano-Plancarte A, Castillo-Hernandez JR. Angiotensin II and the development of insulin resistance: implications for diabetes. Mol Cell Endocrinol.

2009;302:128-139. DOI:10.1016/j.mce.2008.12.011

[60] Leung PS. Mechanisms of protective effects induced by blockade of the renin-angiotensin system: novel role of the pancreatic islet angiotensin-generating system in Type 2 diabetes. Diabet Med. 2007;24:110-116. DOI:10.1111/j.1464-5491.2007.02072.x

[61] Velloso LA, Folli F, Sun XJ, White MF, Saad MJ, Kahn CR. Cross-talk between the insulin and angiotensin signaling systems. Proc Natl Acad Sci U S A. 1996;93:12490-12495. DOI:10.1073/pnas.93.22.12490

[62] Gillespie EL, White CM, Kardas M, Lindberg M, Coleman CI. The impact of ACE inhibitors or angiotensin II type 1 receptor blockers on the development of new-onset type 2 diabetes. Diabetes Care. 2005;28:2261-2266. DOI:10.2337/diacare.28.9.2261

[63] Yang JK, Zhou JB, Xin Z, Zhao L, Yu M, Feng JP, et al. Interactions among related genes of renin-angiotensin system associated with type 2 diabetes. Diabetes Care. 2010;33:2271-2273. DOI:10.2337/dc10-0349

[64] Zhou D, Ruiter R, Zhang J, Zhou M, Liu H, Liu W, et al. Angiotensin-converting enzyme I/D polymorphism is not associated with type 2 diabetes in a Chinese population. J Renin Angiotensin Aldosterone Syst. 2012;13:372-378. DOI:10.1177/1470320311435535

[65] Hansson L, Lindholm LH, Niskanen L, Lanke J, Hedner T, Niklason A, et al. Effect of angiotensin-converting-enzyme inhibition compared with conventional therapy on cardiovascular morbidity and mortality in hypertension: the Captopril Prevention Project (CAPPP) randomised trial. Lancet. 1999;353:611-616. DOI:10.1016/s0140-6736(98)05012-0

[66] Dahlof B, Devereux RB, Kjeldsen SE, Julius S, Beevers G, de Faire U, et al. Cardiovascular morbidity and mortality in the Losartan Intervention For Endpoint reduction in hypertension study (LIFE): a randomised trial against atenolol. Lancet. 2002;359:995-1003. DOI:10.1016/S0140-6736(02)08089-3

[67] Miller JA, Floras JS, Zinman B, Skorecki KL, Logan AG. Effect of hyperglycaemia on arterial pressure, plasma renin activity and renal function in early diabetes. Clin Sci (Lond). 1996;90:189-195. DOI:10.1042/cs0900189

[68] Miller JA. Impact of hyperglycemia on the renin angiotensin system in early human type 1 diabetes mellitus. J Am Soc Nephrol. 1999;10:1778-1785

[69] Leri A, Claudio PP, Li Q, Wang X, Reiss K, Wang S, et al. Stretch-mediated release of angiotensin II induces myocyte apoptosis by activating p53 that enhances the local renin-angiotensin system and decreases the Bcl-2-to-Bax protein ratio in the cell. J Clin Invest. 1998;101:1326-1342. DOI:10.1172/JCI316

[70] Leri A, Fiordaliso F, Setoguchi M, Limana F, Bishopric NH, Kajstura J, et al. Inhibition of p53 function prevents renin-angiotensin system activation and stretch-mediated myocyte apoptosis. Am J Pathol. 2000;157:843-857. DOI:10.1016/S0002-9440(10)64598-1

[71] Fiordaliso F, Leri A, Cesselli D, Limana F, Safai B, Nadal-Ginard B, et al. Hyperglycemia activates p53 and p53-regulated genes leading to myocyte cell death. Diabetes. 2001;50:2363-2375. DOI:10.2337/diabetes.50.10.2363

[72] Purkait P, Halder K, Thakur S, Ghosh Roy A, Raychaudhuri P, Bhattacharya S, et al. Association of angiotensinogen gene SNPs and haplotypes with risk of hypertension in eastern Indian population. Clin

Hypertens. 2017;23:12. DOI:10.1186/s40885-017-0069-x

[73] Lin J, Hu FB, Qi L, Curhan GC. Genetic polymorphisms of angiotensin-2 type 1 receptor and angiotensinogen and risk of renal dysfunction and coronary heart disease in type 2 diabetes mellitus. BMC Nephrol. 2009;10:9. DOI:10.1186/1471-2369-10-9

[74] Chang HR, Cheng CH, Shu KH, Chen CH, Lian JD, Wu MY. Study of the polymorphism of angiotensinogen, anigiotensin-converting enzyme and angiotensin receptor in type II diabetes with end-stage renal disease in Taiwan. J Chin Med Assoc. 2003;66:51-56

[75] Ezzidi I, Mtiraoui N, Kacem M, Chaieb M, Mahjoub T, Almawi WY. Identification of specific angiotensin-converting enzyme variants and haplotypes that confer risk and protection against type 2 diabetic nephropathy. Diabetes Metab Res Rev. 2009;25:717-724. DOI:10.1002/dmrr.1006

[76] van Ittersum FJ, de Man AM, Thijssen S, de Knijff P, Slagboom E, Smulders Y, et al. Genetic polymorphisms of the renin-angiotensin system and complications of insulin-dependent diabetes mellitus. Nephrol Dial Transplant. 2000;15:1000-1007. DOI:10.1093/ndt/15.7.1000

[77] Jeunemaitre X, Soubrier F, Kotelevtsev YV, Lifton RP, Williams CS, Charru A, et al. Molecular basis of human hypertension: role of angiotensinogen. Cell. 1992;71:169-180. DOI:10.1016/0092-8674(92)90275-h

[78] Joyce-Tan SM, Zain SM, Abdul Sattar MZ, Abdullah NA. Renin-Angiotensin System Gene Variants and Type 2 Diabetes Mellitus: Influence of Angiotensinogen. J Diabetes Res. 2016;2016:2161376. DOI:10.1155/2016/2161376

[79] Ichikawa M, Konoshita T, Nakaya T, Yamamoto K, Yamada M, Sato S, et al. Genetic variant of the renin-angiotensin system and prevalence of type 2 diabetes mellitus: a modest but significant effect of aldosterone synthase. Acta Diabetol. 2014;51:595-599. DOI:10.1007/s00592-014-0561-7

[80] Shaikh R, Shahid SM, Mansoor Q, Ismail M, Azhar A. Genetic variants of ACE (Insertion/Deletion) and AGT (M268T) genes in patients with diabetes and nephropathy. J Renin Angiotensin Aldosterone Syst. 2014;15:124-130. DOI:10.1177/1470320313512390

[81] Lee, Sr., Moon JY, Lee SH, Ihm CG, Lee TW, Kim SK, et al. Angiotensinogen polymorphisms and post-transplantation diabetes mellitus in Korean renal transplant subjects. Kidney Blood Press Res. 2013;37:95-102. DOI:10.1159/000343404

[82] Kagami S, Border WA, Miller DE, Noble NA. Angiotensin II stimulates extracellular matrix protein synthesis through induction of transforming growth factor-beta expression in rat glomerular mesangial cells. J Clin Invest. 1994;93:2431-2437. DOI:10.1172/JCI117251

[83] Jaimes EA, Galceran JM, Raij L. Angiotensin II induces superoxide anion production by mesangial cells. Kidney Int. 1998;54:775-784. DOI:10.1046/j.1523-1755.1998.00068.x

[84] Marre M, Jeunemaitre X, Gallois Y, Rodier M, Chatellier G, Sert C, et al. Contribution of genetic polymorphism in the renin-angiotensin system to the development of renal complications in insulin-dependent diabetes: Genetique de la Nephropathie Diabetique (GENEDIAB) study group. J Clin Invest. 1997;99:1585-1595. DOI:10.1172/JCI119321

[85] Chang HF, Hsiao PJ, Hsu YJ, Lin FH, Lin C, Su W, et al. Association between

angiotensin II receptor type 1 A1166C polymorphism and chronic kidney disease. Oncotarget. 2018;9:14444-14455. DOI:10.18632/oncotarget.24469

[86] Ahmad N, Jamal R, Shah SA, Gafor AHA, Murad NAA. Renin-Angiotensin-Aldosterone System Gene Polymorphisms and Type 2 Diabetic Nephropathy in Asian Populations: An Updated Meta-analysis. Curr Diabetes Rev. 2019;15:263-276. DOI:10.2174/15733 99814666180709100411

[87] van Zuydam NR, Ahlqvist E, Sandholm N, Deshmukh H, Rayner NW, Abdalla M, et al. A Genome-Wide Association Study of Diabetic Kidney Disease in Subjects With Type 2 Diabetes. Diabetes. 2018;67:1414-1427. DOI:10.2337/db17-0914

[88] Seaquist ER, Goetz FC, Rich S, Barbosa J. Familial clustering of diabetic kidney disease. Evidence for genetic susceptibility to diabetic nephropathy. N Engl J Med. 1989;320:1161-1165. DOI:10.1056/NEJM198905043201801

[89] Deinum J, Tarnow L, van Gool JM, de Bruin RA, Derkx FH, Schalekamp MA, et al. Plasma renin and prorenin and renin gene variation in patients with insulin-dependent diabetes mellitus and nephropathy. Nephrol Dial Transplant. 1999;14:1904-1911. DOI:10.1093/ndt/14.8.1904

[90] Mansego ML, Redon J, Marin R, Gonzalez-Albert V, Martin-Escudero JC, Fabia MJ, et al. Renin polymorphisms and haplotypes are associated with blood pressure levels and hypertension risk in postmenopausal women. J Hypertens. 2008;26:230-237. DOI:10.1097/HJH.0b013e3282f29865

[91] Hasimu B, Nakayama T, Mizutani Y, Izumi Y, Asai S, Soma M, et al. Haplotype analysis of the human renin gene and essential hypertension. Hypertension. 2003;41:308-312. DOI:10.1161/01.hyp.0000049762.77830.89

[92] Frossard PM, Malloy MJ, Lestringant GG, Kane JP. Haplotypes of the human renin gene associated with essential hypertension and stroke. J Hum Hypertens. 2001;15:49-55. DOI:10.1038/sj.jhh.1001107

[93] Ahmad U, Saleheen D, Bokhari A, Frossard PM. Strong association of a renin intronic dimorphism with essential hypertension. Hypertens Res. 2005;28:339-344. DOI:10.1291/hypres.28.339

[94] Moore N, Dicker P, O'Brien JK, Stojanovic M, Conroy RM, Treumann A, et al. Renin gene polymorphisms and haplotypes, blood pressure, and responses to renin-angiotensin system inhibition. Hypertension. 2007;50:340-347. DOI:10.1161/HYPER TENSIONAHA.106.085563

[95] Sun B, Williams JS, Pojoga L, Chamarthi B, Lasky-Su J, Raby BA, et al. Renin gene polymorphism: its relationship to hypertension, renin levels and vascular responses. J Renin Angiotensin Aldosterone Syst. 2011;12:564-571. DOI:10.1177/1470320311405873

[96] Purkait P, Suthar PC, Purohit VK, Naidu JM, Sarkar BN. Renin-angiotensin-aldosterone system gene polymorphisms in Type 2 diabetic patients among the Mewari population of Rajasthan. Int J Biol Med Res. 2013;4:3128-3134

[97] Berge KE, Berg K. No effect of a BglI polymorphism at the renin (REN) locus on blood pressure level or variability. Clin Genet. 1994;46:436-438. DOI:10.1111/j.1399-0004.1994.tb04413.x

[98] Chiang FT, Hsu KL, Tseng CD, Lo HM, Chern TH, Tseng YZ. Association of the renin gene

polymorphism with essential hypertension in a Chinese population. Clin Genet. 1997;51:370-374. DOI:10.1111/j.1399-0004.1997.tb02493.x

[99] Afruza R, Islam LN, Banerjee S, Hassan MM, Suzuki F, Nabi AN. Renin gene polymorphisms in bangladeshi hypertensive population. J Genomics. 2014;2:45-53. DOI:10.7150/jgen.5193

[100] Prasad P, Tiwari AK, Kumar KM, Ammini AC, Gupta A, Gupta R, et al. Chronic renal insufficiency among Asian Indians with type 2 diabetes: I. Role of RAAS gene polymorphisms. BMC Med Genet. 2006;7:42. DOI:10.1186/1471-2350-7-42

[101] Mtiraoui N, Ezzidi I, Turki A, Chaieb M, Mahjoub T, Almawi WY. Renin-angiotensin-aldosterone system ps and haplotypes affect the susceptibility to nephropathy in type 2 diabetes patients. J Renin Angiotensin Aldosterone Syst. 2011;12:572-580. DOI:10.1177/1470320310396542

[102] Purkait P, Halder K, Naidu JM, Sarkar B. Genetic variants of renin on the prevalence of diabetic nephropathy. Vessel Plus. 2018;2:19. DOI:10.20517/2574-1209.2018.16

[103] Germain S, Philippe J, Fuchs S, Lengronne A, Corvol P, Pinet F. Regulation of human renin secretion and gene transcription in Calu-6 cells. FEBS Lett. 1997;407:177-183. DOI:10.1016/s0014-5793(97)00332-3

[104] Russo P, Siani A, Venezia A, Iacone R, Russo O, Barba G, et al. Interaction between the C(−344)T polymorphism of CYP11B2 and age in the regulation of blood pressure and plasma aldosterone levels: cross-sectional and longitudinal findings of the Olivetti Prospective Heart Study. J Hypertens. 2002;20:1785-1792. DOI:10.1097/00004872-200209000-00023

[105] Keavney B, Mayosi B, Gaukrodger N, Imrie H, Baker M, Fraser R, et al. Genetic variation at the locus encompassing 11-beta hydroxylase and aldosterone synthase accounts for heritability in cortisol precursor (11-deoxycortisol) urinary metabolite excretion. J Clin Endocrinol Metab. 2005;90:1072-1077. DOI:10.1210/jc.2004-0870

[106] Kumar NN, Benjafield AV, Lin RC, Wang WY, Stowasser M, Morris BJ. Haplotype analysis of aldosterone synthase gene (CYP11B2) polymorphisms shows association with essential hypertension. J Hypertens. 2003;21:1331-1337. DOI:10.1097/00004872-200307000-00022

[107] Tamaki S, Iwai N, Tsujita Y, Kinoshita M. Genetic polymorphism of CYP11B2 gene and hypertension in Japanese. Hypertension. 1999;33:266-270. DOI:10.1161/01.hyp.33.1.266

[108] Saidi S, Mahjoub T, Almawi WY. Aldosterone synthase gene (CYP11B2) promoter polymorphism as a risk factor for ischaemic stroke in Tunisian Arabs. J Renin Angiotensin Aldosterone Syst. 2010;11:180-186. DOI:10.1177/1470320309360816

[109] Lovati E, Richard A, Frey BM, Frey FJ, Ferrari P. Genetic polymorphisms of the renin-angiotensin-aldosterone system in end-stage renal disease. Kidney Int. 2001;60:46-54. DOI:10.1046/j.1523-1755.2001.00769.x

[110] Song J, Narita I, Goto S, Saito N, Omori K, Sato F, et al. Gender specific association of aldosterone synthase gene polymorphism with renal survival in patients with IgA nephropathy. J Med Genet. 2003;40:372-376. DOI:10.1136/jmg.40.5.372

[111] Lee JE, Bae SY, Kim JY, Pyo HJ, Western Dialysis Physician A, Kwon YJ.

Aldosterone Synthase Gene (CYP11B2) Polymorphism in Korean End-Stage Renal Disease Patients on Hemodialysis. Electrolyte Blood Press. 2009;7:67-72. DOI:10.5049/EBP.2009.7.2.67

[112] Xu H, Wang X, Liu M, Shao X, He X. Association of aldosterone synthase (CYP11B2) -344 T/C polymorphism with diabetic nephropathy: A meta-analysis. J Renin Angiotensin Aldosterone Syst. 2016;17:1470320316633896. DOI:10.1177/1470320316633896

[113] Purkait P, Raychodhury P, Bandhyopadhya S, Naidu JM, Sarkar BN. Analysis of aldosterone synthase gene promoter (−344 C>T) polymorphism in Indian diabetic nephropathy patients. J Diabetes Metab. 2013;4:271

[114] Phillips T. The role of methylation in gene expression. Nat Educ. 2008;1:116

[115] Bassett MH, Zhang Y, Clyne C, White PC, Rainey WE. Differential regulation of aldosterone synthase and 11beta-hydroxylase transcription by steroidogenic factor-1. J Mol Endocrinol. 2002;28:125-135. DOI:10.1677/jme.0.0280125

[116] Esparza-Romero J, Valencia ME, Urquidez-Romero R, Chaudhari LS, Hanson RL, Knowler WC, et al. Environmentally Driven Increases in Type 2 Diabetes and Obesity in Pima Indians and Non-Pimas in Mexico Over a 15-Year Period: The Maycoba Project. Diabetes Care. 2015;38:2075-2082. DOI:10.2337/dc15-0089

[117] Florez JC, Hirschhorn J, Altshuler D. The inherited basis of diabetes mellitus: implications for the genetic analysis of complex traits. Annu Rev Genomics Hum Genet. 2003;4:257-291. DOI:10.1146/annurev.genom.4.070802.110436

[118] Hansen L. Candidate genes and late-onset type 2 diabetes mellitus. Susceptibility genes or common polymorphisms? Dan Med Bull. 2003;50:320-346

[119] Gloyn AL. The search for type 2 diabetes genes. Ageing Res Rev. 2003;2:111-127. DOI:10.1016/s1568-1637(02)00061-2

[120] Redondo MJ, Jeffrey J, Fain PR, Eisenbarth GS, Orban T. Concordance for islet autoimmunity among monozygotic twins. N Engl J Med. 2008;359:2849-2850. DOI:10.1056/NEJMc0805398

[121] Jerram ST, Leslie RD. The Genetic Architecture of Type 1 Diabetes. Genes (Basel). 2017;8. DOI:10.3390/genes8080209

[122] Meigs JB, Cupples LA, Wilson PW. Parental transmission of type 2 diabetes: the Framingham Offspring Study. Diabetes. 2000;49:2201-2207. DOI:10.2337/diabetes.49.12.2201

[123] Poulsen P, Kyvik KO, Vaag A, Beck-Nielsen H. Heritability of type II (non-insulin-dependent) diabetes mellitus and abnormal glucose tolerance--a population-based twin study. Diabetologia. 1999;42:139-145. DOI:10.1007/s001250051131

[124] Tillil H, Kobberling J. Age-corrected empirical genetic risk estimates for first-degree relatives of IDDM patients. Diabetes. 1987;36:93-99. DOI:10.2337/diab.36.1.93

[125] Kaprio J, Tuomilehto J, Koskenvuo M, Romanov K, Reunanen A, Eriksson J, et al. Concordance for type 1 (insulin-dependent) and type 2 (non-insulin-dependent) diabetes mellitus in a population-based cohort of twins in Finland. Diabetologia. 1992;35:1060-1067. DOI:10.1007/BF02221682

[126] Fuchsberger C, Flannick J, Teslovich TM, Mahajan A, Agarwala V,

Gaulton KJ, et al. The genetic architecture of type 2 diabetes. Nature. 2016;536:41-47. DOI:10.1038/nature18642

[127] Fajans SS, Bell GI, Polonsky KS. Molecular mechanisms and clinical pathophysiology of maturity-onset diabetes of the young. N Engl J Med. 2001;345:971-980. DOI:10.1056/NEJMra002168

[128] Barroso I. Genetics of Type 2 diabetes. Diabet Med. 2005;22:517-535. DOI:10.1111/j.1464-5491.2005.01550.x

[129] Vaxillaire M, Froguel P. Monogenic diabetes in the young, pharmacogenetics and relevance to multifactorial forms of type 2 diabetes. Endocr Rev. 2008;29:254-264. DOI:10.1210/er.2007-0024

[130] Reynisdottir I, Thorleifsson G, Benediktsson R, Sigurdsson G, Emilsson V, Einarsdottir AS, et al. Localization of a susceptibility gene for type 2 diabetes to chromosome 5q34-q35.2. Am J Hum Genet. 2003;73:323-335. DOI:10.1086/377139

[131] Hunt KJ, Lehman DM, Arya R, Fowler S, Leach RJ, Goring HH, et al. Genome-wide linkage analyses of type 2 diabetes in Mexican Americans: the San Antonio Family Diabetes/Gallbladder Study. Diabetes. 2005;54:2655-2662. DOI:10.2337/diabetes.54.9.2655

[132] Grant SF, Thorleifsson G, Reynisdottir I, Benediktsson R, Manolescu A, Sainz J, et al. Variant of transcription factor 7-like 2 (TCF7L2) gene confers risk of type 2 diabetes. Nat Genet. 2006;38:320-323. DOI:10.1038/ng1732

[133] Das SK, Elbein SC. The Genetic Basis of Type 2 Diabetes. Cellscience. 2006;2:100-131. DOI:10.1901/jaba.2006.2-100

[134] Malhotra A, Igo RP, Jr., Thameem F, Kao WH, Abboud HE, Adler SG, et al. Genome-wide linkage scans for type 2 diabetes mellitus in four ethnically diverse populations-significant evidence for linkage on chromosome 4q in African Americans: the Family Investigation of Nephropathy and Diabetes Research Group. Diabetes Metab Res Rev. 2009;25:740-747. DOI:10.1002/dmrr.1031

[135] Horikawa Y, Oda N, Cox NJ, Li X, Orho-Melander M, Hara M, et al. Genetic variation in the gene encoding calpain-10 is associated with type 2 diabetes mellitus. Nat Genet. 2000;26:163-175. DOI:10.1038/79876

[136] Meyre D, Bouatia-Naji N, Tounian A, Samson C, Lecoeur C, Vatin V, et al. Variants of ENPP1 are associated with childhood and adult obesity and increase the risk of glucose intolerance and type 2 diabetes. Nat Genet. 2005;37:863-867. DOI:10.1038/ng1604

[137] Love-Gregory LD, Wasson J, Ma J, Jin CH, Glaser B, Suarez BK, et al. A common polymorphism in the upstream promoter region of the hepatocyte nuclear factor-4 alpha gene on chromosome 20q is associated with type 2 diabetes and appears to contribute to the evidence for linkage in an ashkenazi jewish population. Diabetes. 2004;53:1134-1140. DOI:10.2337/diabetes.53.4.1134

[138] Silander K, Mohlke KL, Scott LJ, Peck EC, Hollstein P, Skol AD, et al. Genetic variation near the hepatocyte nuclear factor-4 alpha gene predicts susceptibility to type 2 diabetes. Diabetes. 2004;53:1141-1149. DOI:10.2337/diabetes.53.4.1141

[139] Vozarova de Courten B, Hanson RL, Funahashi T, Lindsay RS, Matsuzawa Y, Tanaka S, et al. Common Polymorphisms in the Adiponectin Gene ACDC Are Not Associated With Diabetes in Pima Indians. Diabetes.

2005;54:284-289. DOI:10.2337/diabetes.54.1.284

[140] Tsuchiya T, Schwarz PE, Bosque-Plata LD, Geoffrey Hayes M, Dina C, Froguel P, et al. Association of the calpain-10 gene with type 2 diabetes in Europeans: results of pooled and meta-analyses. Mol Genet Metab. 2006;89:174-184. DOI:10.1016/j.ymgme.2006.05.013

[141] Maddux BA, Goldfine ID. Membrane glycoprotein PC-1 inhibition of insulin receptor function occurs via direct interaction with the receptor alpha-subunit. Diabetes. 2000;49:13-19. DOI:10.2337/diabetes.49.1.13

[142] Ihara A, Yamagata K, Nammo T, Miura A, Yuan M, Tanaka T, et al. Functional characterization of the HNF4alpha isoform (HNF4alpha8) expressed in pancreatic beta-cells. Biochem Biophys Res Commun. 2005;329:984-990. DOI:10.1016/j.bbrc.2005.02.072

[143] Rhee J, Inoue Y, Yoon JC, Puigserver P, Fan M, Gonzalez FJ, et al. Regulation of hepatic fasting response by PPARgamma coactivator-1alpha (PGC-1): requirement for hepatocyte nuclear factor 4alpha in gluconeogenesis. Proc Natl Acad Sci U S A. 2003;100:4012-4017. DOI:10.1073/pnas.0730870100

[144] Yamagata K, Furuta H, Oda N, Kaisaki PJ, Menzel S, Cox NJ, et al. Mutations in the hepatocyte nuclear factor-4alpha gene in maturity-onset diabetes of the young (MODY1). Nature. 1996;384:458-460. DOI:10.1038/384458a0

[145] Cho YS, Chen CH, Hu C, Long J, Ong RT, Sim X, et al. Meta-analysis of genome-wide association studies identifies eight new loci for type 2 diabetes in east Asians. Nat Genet. 2011;44:67-72. DOI:10.1038/ng.1019

[146] Jafar-Mohammadi B, Groves CJ, Gjesing AP, Herrera BM, Winckler W, Stringham HM, et al. A role for coding functional variants in HNF4A in type 2 diabetes susceptibility. Diabetologia. 2011;54:111-119. DOI:10.1007/s00125-010-1916-4

[147] Kooner JS, Saleheen D, Sim X, Sehmi J, Zhang W, Frossard P, et al. Genome-wide association study in individuals of South Asian ancestry identifies six new type 2 diabetes susceptibility loci. Nat Genet. 2011;43:984-989. DOI:10.1038/ng.921

[148] Weyer C, Funahashi T, Tanaka S, Hotta K, Matsuzawa Y, Pratley RE, et al. Hypoadiponectinemia in obesity and type 2 diabetes: close association with insulin resistance and hyperinsulinemia. J Clin Endocrinol Metab. 2001;86:1930-1935. DOI:10.1210/jcem.86.5.7463

[149] Arita Y, Kihara S, Ouchi N, Takahashi M, Maeda K, Miyagawa J, et al. Paradoxical decrease of an adipose-specific protein, adiponectin, in obesity. Biochem Biophys Res Commun. 1999;257:79-83. DOI:10.1006/bbrc.1999.0255

[150] Hara K, Boutin P, Mori Y, Tobe K, Dina C, Yasuda K, et al. Genetic variation in the gene encoding adiponectin is associated with an increased risk of type 2 diabetes in the Japanese population. Diabetes. 2002;51:536-540. DOI:10.2337/diabetes.51.2.536

[151] Stumvoll M, Tschritter O, Fritsche A, Staiger H, Renn W, Weisser M, et al. Association of the T-G polymorphism in adiponectin (exon 2) with obesity and insulin sensitivity: interaction with family history of type 2 diabetes. Diabetes. 2002;51:37-41. DOI:10.2337/diabetes.51.1.37

[152] Liu Z, Habener JF. Wnt signaling in pancreatic islets. Adv Exp Med Biol.

2010;654:391-419. DOI:10.1007/978-90-481-3271-3_17

[153] Lawlor N, George J, Bolisetty M, Kursawe R, Sun L, Sivakamasundari V, et al. Single-cell transcriptomes identify human islet cell signatures and reveal cell-type-specific expression changes in type 2 diabetes. Genome Res. 2017;27:208-222. DOI:10.1101/gr.212720.116

[154] Diabetes Genetics Initiative of Broad Institute of H, Mit LU, Novartis Institutes of BioMedical R, Saxena R, Voight BF, Lyssenko V, et al. Genome-wide association analysis identifies loci for type 2 diabetes and triglyceride levels. Science. 2007;316:1331-1336. DOI:10.1126/science.1142358

[155] Scott LJ, Mohlke KL, Bonnycastle LL, Willer CJ, Li Y, Duren WL, et al. A genome-wide association study of type 2 diabetes in Finns detects multiple susceptibility variants. Science. 2007;316:1341-1345. DOI:10.1126/science.1142382

[156] Groenewoud MJ, Dekker JM, Fritsche A, Reiling E, Nijpels G, Heine RJ, et al. Variants of CDKAL1 and IGF2BP2 affect first-phase insulin secretion during hyperglycaemic clamps. Diabetologia. 2008;51:1659-1663. DOI:10.1007/s00125-008-1083-z

[157] Louveau I, Gondret F. Regulation of development and metabolism of adipose tissue by growth hormone and the insulin-like growth factor system. Domest Anim Endocrinol. 2004;27:241-255. DOI:10.1016/j.domaniend.2004.06.004

[158] Lasram K, Ben Halim N, Benrahma H, Mediene-Benchekor S, Arfa I, Hsouna S, et al. Contribution of CDKAL1 rs7756992 and IGF2BP2 rs4402960 polymorphisms in type 2 diabetes, diabetic complications, obesity risk and hypertension in the Tunisian population. J Diabetes. 2015;7:102-113. DOI:10.1111/1753-0407.12147

[159] McTaggart JS, Clark RH, Ashcroft FM. The role of the KATP channel in glucose homeostasis in health and disease: more than meets the islet. J Physiol. 2010;588:3201-3209. DOI:10.1113/jphysiol.2010.191767

[160] Gloyn AL, Weedon MN, Owen KR, Turner MJ, Knight BA, Hitman G, et al. Large-scale association studies of variants in genes encoding the pancreatic beta-cell KATP channel subunits Kir6.2 (KCNJ11) and SUR1 (ABCC8) confirm that the KCNJ11 E23K variant is associated with type 2 diabetes. Diabetes. 2003;52:568-572. DOI:10.2337/diabetes.52.2.568

[161] Imamura M, Maeda S. Genetics of type 2 diabetes: the GWAS era and future perspectives [Review]. Endocr J. 2011;58:723-739. DOI:10.1507/endocrj.ej11-0113

[162] International HapMap C. A haplotype map of the human genome. Nature. 2005;437:1299-1320. DOI:10.1038/nature04226

[163] McCarthy MI, Abecasis GR, Cardon LR, Goldstein DB, Little J, Ioannidis JP, et al. Genome-wide association studies for complex traits: consensus, uncertainty and challenges. Nat Rev Genet. 2008;9:356-369. DOI:10.1038/nrg2344

[164] Small KS, Todorcevic M, Civelek M, El-Sayed Moustafa JS, Wang X, Simon MM, et al. Regulatory variants at KLF14 influence type 2 diabetes risk via a female-specific effect on adipocyte size and body composition. Nat Genet. 2018;50:572-580. DOI:10.1038/s41588-018-0088-x

[165] Kong A, Steinthorsdottir V, Masson G, Thorleifsson G, Sulem P, Besenbacher S, et al. Parental origin of

sequence variants associated with complex diseases. Nature. 2009;462:868-874. DOI:10.1038/nature08625

[166] Chen YH, Xu SJ, Bendahhou S, Wang XL, Wang Y, Xu WY, et al. KCNQ1 gain-of-function mutation in familial atrial fibrillation. Science. 2003;299:251-254. DOI:10.1126/science.1077771

[167] Tan JT, Nurbaya S, Gardner D, Ye S, Tai ES, Ng DP. Genetic variation in KCNQ1 associates with fasting glucose and beta-cell function: a study of 3,734 subjects comprising three ethnicities living in Singapore. Diabetes. 2009;58:1445-1449. DOI:10.2337/db08-1138

[168] Liu Y, Zhou DZ, Zhang D, Chen Z, Zhao T, Zhang Z, et al. Variants in KCNQ1 are associated with susceptibility to type 2 diabetes in the population of mainland China. Diabetologia. 2009;52:1315-1321. DOI:10.1007/s00125-009-1375-y

[169] Fukuda H, Imamura M, Tanaka Y, Iwata M, Hirose H, Kaku K, et al. A single nucleotide polymorphism within DUSP9 is associated with susceptibility to type 2 diabetes in a Japanese population. PLoS One. 2012;7:e46263. DOI:10.1371/journal.pone.0046263

[170] Rees SD, Hydrie MZ, Shera AS, Kumar S, O'Hare JP, Barnett AH, et al. Replication of 13 genome-wide association (GWA)-validated risk variants for type 2 diabetes in Pakistani populations. Diabetologia. 2011;54:1368-1374. DOI:10.1007/s00125-011-2063-2

[171] Morris AP, Voight BF, Teslovich TM, Ferreira T, Segre AV, Steinthorsdottir V, et al. Large-scale association analysis provides insights into the genetic architecture and pathophysiology of type 2 diabetes. Nat Genet. 2012;44:981-990. DOI:10.1038/ng.2383

[172] Hertel JK, Johansson S, Sonestedt E, Jonsson A, Lie RT, Platou CG, et al. FTO, type 2 diabetes, and weight gain throughout adult life: a meta-analysis of 41,504 subjects from the Scandinavian HUNT, MDC, and MPP studies. Diabetes. 2011;60:1637-1644. DOI:10.2337/db10-1340

[173] Ragvin A, Moro E, Fredman D, Navratilova P, Drivenes O, Engstrom PG, et al. Long-range gene regulation links genomic type 2 diabetes and obesity risk regions to HHEX, SOX4, and IRX3. Proc Natl Acad Sci U S A. 2010;107:775-780. DOI:10.1073/pnas.0911591107

[174] Cauchi S, Meyre D, Durand E, Proenca C, Marre M, Hadjadj S, et al. Post genome-wide association studies of novel genes associated with type 2 diabetes show gene-gene interaction and high predictive value. PLoS One. 2008;3:e2031. DOI:10.1371/journal.pone.0002031

[175] Huang Q, Yin JY, Dai XP, Pei Q, Dong M, Zhou ZG, et al. IGF2BP2 variations influence repaglinide response and risk of type 2 diabetes in Chinese population. Acta Pharmacol Sin. 2010;31:709-717. DOI:10.1038/aps.2010.47

[176] Gamboa-Melendez MA, Huerta-Chagoya A, Moreno-Macias H, Vazquez-Cardenas P, Ordonez-Sanchez ML, Rodriguez-Guillen R, et al. Contribution of common genetic variation to the risk of type 2 diabetes in the Mexican Mestizo population. Diabetes. 2012;61:3314-3321. DOI:10.2337/db11-0550

[177] Xiao S, Zeng X, Fan Y, Su Y, Ma Q, Zhu J, et al. Gene Polymorphism Association with Type 2 Diabetes and Related Gene-Gene and

Gene-Environment Interactions in a Uyghur Population. Med Sci Monit. 2016;22:474-487. DOI:10.12659/msm.895347

[178] Cunnington MS, Santibanez Koref M, Mayosi BM, Burn J, Keavney B. Chromosome 9p21 SNPs Associated with Multiple Disease Phenotypes Correlate with ANRIL Expression. PLoS Genet. 2010;6:e1000899. DOI:10.1371/journal.pgen.1000899

[179] Horikoshi M, Hara K, Ito C, Shojima N, Nagai R, Ueki K, et al. Variations in the HHEX gene are associated with increased risk of type 2 diabetes in the Japanese population. Diabetologia. 2007;50:2461-2466. DOI:10.1007/s00125-007-0827-5

[180] Herder C, Rathmann W, Strassburger K, Finner H, Grallert H, Huth C, et al. Variants of the PPARG, IGF2BP2, CDKAL1, HHEX, and TCF7L2 genes confer risk of type 2 diabetes independently of BMI in the German KORA studies. Horm Metab Res. 2008;40:722-726. DOI:10.1055/s-2008-1078730

[181] Lee YH, Kang ES, Kim SH, Han SJ, Kim CH, Kim HJ, et al. Association between polymorphisms in SLC30A8, HHEX, CDKN2A/B, IGF2BP2, FTO, WFS1, CDKAL1, KCNQ1 and type 2 diabetes in the Korean population. J Hum Genet. 2008;53:991-998. DOI:10.1007/s10038-008-0341-8

[182] Bhowmick A, Sarkar P, Baruah MP, Bodhini D, Radha V, Mohan V, et al. Association of SLC30A8, CDKAL1, TCF7L2 and HHEX Gene Polymorphisms with Type 2 Diabetes in the Population of North East India. Cytology and Genetics. 2020;54:165-172. DOI:10.3103/s0095452720020036

[183] Sladek R, Rocheleau G, Rung J, Dina C, Shen L, Serre D, et al. A genome-wide association study identifies novel risk loci for type 2 diabetes. Nature. 2007;445:881-885. DOI:10.1038/nature05616

[184] Dupuis J, Langenberg C, Prokopenko I, Saxena R, Soranzo N, Jackson AU, et al. New genetic loci implicated in fasting glucose homeostasis and their impact on type 2 diabetes risk. Nat Genet. 2010;42:105-116. DOI:10.1038/ng.520

[185] Strawbridge RJ, Dupuis J, Prokopenko I, Barker A, Ahlqvist E, Rybin D, et al. Genome-wide association identifies nine common variants associated with fasting proinsulin levels and provides new insights into the pathophysiology of type 2 diabetes. Diabetes. 2011;60:2624-2634. DOI:10.2337/db11-0415

[186] Flannick J, Thorleifsson G, Beer NL, Jacobs SB, Grarup N, Burtt NP, et al. Loss-of-function mutations in SLC30A8 protect against type 2 diabetes. Nat Genet. 2014;46:357-363. DOI:10.1038/ng.2915

[187] Xu J, Wang J, Chen B. SLC30A8 (ZnT8) variations and type 2 diabetes in the Chinese Han population. Genet Mol Res. 2012;11:1592-1598. DOI:10.4238/2012.May.24.1

[188] Voight BF, Scott LJ, Steinthorsdottir V, Morris AP, Dina C, Welch RP, et al. Twelve type 2 diabetes susceptibility loci identified through large-scale association analysis. Nat Genet. 2010;42:579-589. DOI:10.1038/ng.609

[189] Saxena R, Hivert MF, Langenberg C, Tanaka T, Pankow JS, Vollenweider P, et al. Genetic variation in GIPR influences the glucose and insulin responses to an oral glucose challenge. Nat Genet. 2010;42:142-148. DOI:10.1038/ng.521

[190] Zeggini E, Scott LJ, Saxena R, Voight BF, Marchini JL, Hu T, et al. Meta-analysis of genome-wide

association data and large-scale replication identifies additional susceptibility loci for type 2 diabetes. Nat Genet. 2008;40:638-645. DOI:10.1038/ng.120

[191] Winckler W, Weedon MN, Graham RR, McCarroll SA, Purcell S, Almgren P, et al. Evaluation of common variants in the six known maturity-onset diabetes of the young (MODY) genes for association with type 2 diabetes. Diabetes. 2007;56:685-693. DOI:10.2337/db06-0202

[192] Ouederni TB, Sanchez-Corona J, Flores Martinez SE, Ben Maiz H, Skhiri HA, Abid HK, et al. The G1057D polymorphism of IRS-2 gene is not associated with type 2 diabetes and obese patients among ethnic groups in Tunisian population. Clin Biochem. 2009;42:1169-1173. DOI:10.1016/j.clinbiochem.2009.03.018

[193] Bappy HMJA, Goswami A, Huda N, Hosen MI, Nabi AHMN. Gender specific association of missense variant rs1805097 of IRS-2 and noncoding variant rs841853 of GLUT-1 genes with susceptibility to type 2 diabetes in Bangladeshi population. Gene Reports. 2020;21:100866. DOI:10.1016/j.genrep.2020.100866

[194] Huda N, Hosen MI, Yasmin T, Sarkar PK, Hasan A, Nabi A. Genetic variation of the transcription factor GATA3, not STAT4, is associated with the risk of type 2 diabetes in the Bangladeshi population. PLoS One. 2018;13:e0198507. DOI:10.1371/journal.pone.0198507

[195] Scott RA, Scott LJ, Magi R, Marullo L, Gaulton KJ, Kaakinen M, et al. An Expanded Genome-Wide Association Study of Type 2 Diabetes in Europeans. Diabetes. 2017;66:2888-2902. DOI:10.2337/db16-1253

[196] Rung J, Cauchi S, Albrechtsen A, Shen L, Rocheleau G, Cavalcanti-Proenca C, et al. Genetic variant near IRS1 is associated with type 2 diabetes, insulin resistance and hyperinsulinemia. Nat Genet. 2009;41:1110-1115. DOI:10.1038/ng.443

[197] Goswami A, Huda N, Yasmin T, Hosen MI, Hasan A, Nabi A. Association study of leukocyte telomere length and genetic polymorphism within hTERT promoter with type 2 diabetes in Bangladeshi population. Mol Biol Rep. 2021. DOI:10.1007/s11033-020-06045-7

[198] Jeunemaitre X. Genetics of the human renin angiotensin system. J Mol Med (Berl). 2008;86:637-641. DOI:10.1007/s00109-008-0344-0

[199] Sethi AA, Nordestgaard BG, Gronholdt ML, Steffensen R, Jensen G, Tybjaerg-Hansen A. Angiotensinogen single nucleotide polymorphisms, elevated blood pressure, and risk of cardiovascular disease. Hypertension. 2003;41:1202-1211. DOI:10.1161/01.HYP.0000072334.34433.17

[200] Sethi AA, Nordestgaard BG, Tybjaerg-Hansen A. Angiotensinogen gene polymorphism, plasma angiotensinogen, and risk of hypertension and ischemic heart disease: a meta-analysis. Arterioscler Thromb Vasc Biol. 2003;23:1269-1275. DOI:10.1161/01.ATV.0000079007.40884.5C

[201] Watkins WS, Rohrwasser A, Peiffer A, Leppert MF, Lalouel JM, Jorde LB. AGT genetic variation, plasma AGT, and blood pressure: An analysis of the Utah Genetic Reference Project pedigrees. Am J Hypertens. 2010;23:917-923. DOI:10.1038/ajh.2010.83

[202] Sato N, Katsuya T, Nakagawa T, Ishikawa K, Fu Y, Asai T, et al. Nine polymorphisms of angiotensinogen gene in the susceptibility to essential hypertension. Life Sci. 2000;68:259-272. DOI:10.1016/s0024-3205(00)00795-5

[203] Fejerman L, Wu X, Adeyemo A, Luke A, Zhu X, Hicks C, et al. The effect of genetic variation in angiotensinogen on serum levels and blood pressure: a comparison of Nigerians and US blacks. J Hum Hypertens. 2006;20:882-887. DOI:10.1038/sj.jhh.1002077

[204] Ramachandran V, Ismail P, Stanslas J, Shamsudin N. Analysis of renin-angiotensin aldosterone system gene polymorphisms in Malaysian essential hypertensive and type 2 diabetic subjects. Cardiovasc Diabetol. 2009;8:11. DOI:10.1186/1475-2840-8-11

[205] Ji L, Cai X, Zhang L, Fei L, Wang L, Su J, et al. Association between polymorphisms in the renin-angiotensin-aldosterone system genes and essential hypertension in the Han Chinese population. PLoS One. 2013;8:e72701. DOI:10.1371/journal.pone.0072701

[206] Chamarthi B, Williams JS, Williams GH. A mechanism for salt-sensitive hypertension: abnormal dietary sodium-mediated vascular response to angiotensin-II. J Hypertens. 2010;28:1020-1026. DOI:10.1097/HJH.0b013e3283375974

[207] Rossier BC, Bochud M, Devuyst O. The Hypertension Pandemic: An Evolutionary Perspective. Physiology (Bethesda). 2017;32:112-125. DOI:10.1152/physiol.00026.2016

[208] Ji LD, Li JY, Yao BB, Cai XB, Shen QJ, Xu J. Are genetic polymorphisms in the renin-angiotensin-aldosterone system associated with essential hypertension? Evidence from genome-wide association studies. J Hum Hypertens. 2017;31:695-698. DOI:10.1038/jhh.2017.29

[209] Lip S, Padmanabhan S. Genomics of Blood Pressure and Hypertension: Extending the Mosaic Theory Toward Stratification. Can J Cardiol. 2020;36:694-705. DOI:10.1016/j.cjca.2020.03.001

[210] Devaux CA, Rolain JM, Raoult D. ACE2 receptor polymorphism: Susceptibility to SARS-CoV-2, hypertension, multi-organ failure, and COVID-19 disease outcome. J Microbiol Immunol Infect. 2020;53:425-435. DOI:10.1016/j.jmii.2020.04.015

[211] Akther J, Das A, Rahman MA, Saha SK, Hosen MI, Ebihara A, et al. Non-coding single-nucleotide variants of renin and the (pro)renin receptor are associated with polygenic diseases in a Bangladeshi population. Biochem Genet. 2021. DOI:10.1007/s10528-021-10049-8

Section 2

The Renin-Angiotensin Aldosterone System in Various Disorders

Chapter 3

Role of the Renin-Angiotensin-Aldosterone System in Various Disease Processes: An Overview

Volkan Gelen, Abdulsamed Kükürt and Emin Şengül

Abstract

The renin-angiotensin-aldosterone system is a physiological system that plays an important role in the regulation of blood pressure and body water-electrolyte balance, in which the kidney, liver and lungs play a role in its activation. This system comes into play in various diseases such as the cardiovascular, renal, pulmonary and nervous system where blood pressure and fluid-electrolyte balance may change. The purpose of this study, which is presented in line with this information, is to explain the working principle of this system, how this system is activated, how it comes into play in the mentioned diseases, and what kind of results occur.

Keywords: Renin, angiotensin, aldosterone, ACE2, hypertension, pulmonary diseases, renal diseases, neurodegenerative diseases, AngII, Covid-19

1. Introduction

The renin-angiotensin-aldosterone system (RAAS) is a powerful system that regulates fluid-electrolyte balance and systemic blood pressure. First, it has been stated that it is a hormonal and peptidergic endocrine system that regulates blood pressure and fluid-electrolyte balance [1, 2]. Until recently, RAAS was known only as an endocrine system that regulates blood pressure and fluid-electrolyte balance, but now it is noted that this system is not only found in circulation but also locally in organ systems, and also has autocrine-paracrine functions [3].

There are some components of RAAS responsible for these effects. One of these components, renin, is synthesised as prorenin from the juxtaglomerular apparatus, which is also found in kidney efferent arterioles. The protein is converted to active renin, stored in secretory granules and released into the circulation when necessary [4]. The release of renin, a proteolytic enzyme, is triggered by many physiological stimuli, including prostacyclins (PGI2), such as stimulation of macula densa in the distal tubule with low Na + concentration, reduction of arterial pressure, renal sympathetic nerve activation and stimulation of β1-receptors [5]. Circulating renin provides the formation of Angiotensin I (AngI) from angiotensinogen, most of which is synthesised from the liver [6]. AngI is converted to Angiotensin II (AngII) by Angiotensin-converting enzyme (ACE), a membrane-bound metalloproteinase found in high amounts on pulmonary vascular endothelial cell surfaces (**Figure 1**) [5, 7].

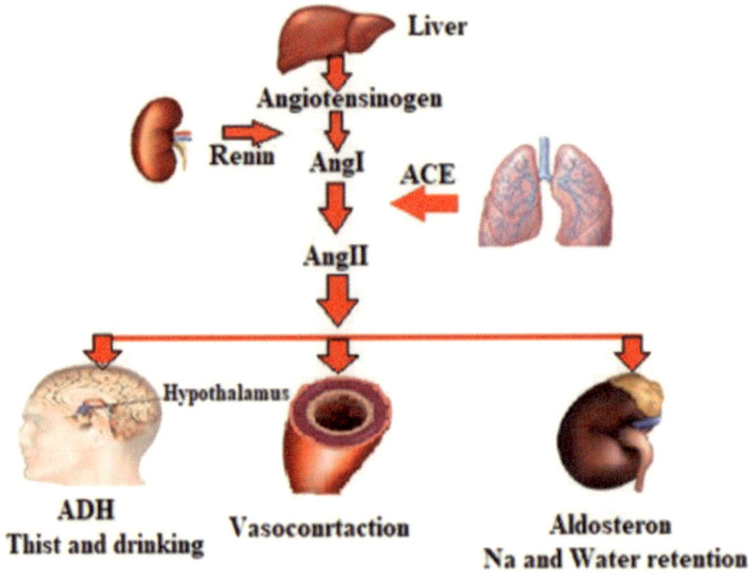

Figure 1.
Renin-angiotensin-aldosterone system and effects.

ACE, a member of the zinc metallopeptidase class, had two main roles in metabolism. It takes part in the RAAS system and the kinin-kallikrein system (KKS). Another task is to inactivate substance P and neurokines [8, 9]. ACE has two forms in endothelial and epithelial cells and male spermatid. Its form in endothelial and epithelial cells is called "somatic form" (sACE), and the form found in spermatids is called "germinal form" (gACE) [10]. The primary structure of these two forms is different from each other. While sACE has two active sites with different catalytic properties, gACE has only one active [11]. ACE has another mammalian homologue named angiotensin-converting enzyme 2 (ACE2) [12]. Although ACE2 has carboxypeptidase activity like ACE, it cleaves an amino acid unlike ACE and its most important substrates are AngI and AngII [13].

In the body, AngII has many roles such as increasing blood pressure by direct contraction of vascular smooth muscles, increasing myocardial contractility, water and salt retention by stimulating aldosterone release from the adrenals, stimulation of catecholamine release from sympathetic nerve endings, cell growth and proliferation [14, 15]. It turns out that AngII can be generated locally in many tissues, including the brain, independent of circulating components [16]. AngII acts by binding to receptors in the protein structure on the plasma membranes of different tissues. These receptors are termed AngII type 1 (AT1R) and AngII type 2 (AT2R) receptors [17]. Changes in the balance of RAAS have been reported to have direct or indirect effects with cardiovascular system diseases, lung diseases, nervous system diseases and kidney diseases. Therefore, this section describes the mechanism of action of RAAS and the relationship of RAAS components with these diseases.

2. The role of RAAS in cardiovascular disease

2.1 Heart failure and myocardial infarction

Ang II has a role in a variety of cardiac dysfunctions, including hypertrophy, arrhythmia, and ventricular dysfunction [18, 19]. Inability to pump enough blood to the body due to insufficient heart functions due to various reasons is known as heart failure. When looking at the role of RAAS in the case of heart failure, RAAS activation can occur when hypertrophy occurs in the heart muscle cells. This causes fluid retention in the body and peripheral vasoconstriction, resulting in cardiac overload and heart failure [20]. RAAS activation increases in heart rate and contractility, thus reducing coronary blood flow [21]. Experimental studies have shown that plasma renin activity increases in acute heart failure. Also, it was determined that plasma renin activity was normal in the compensated phase of chronic heart failure, and this shows that RAAS is associated with heart failure [22]. It has also been determined that when myocardial cells are exposed to excessive AngII and aldosterone, fibrosis is formed. This again shows that RAAS plays an important role in myocardial heart disease. It was determined that AT-1 receptor expression affected by AngII decreased in decompensated heart failure, while AT-2 receptors remained unchanged [23]. It has also been determined that ACE inhibitors play an important role in heart failure. It has been reported that ACE inhibitors are beneficial, especially in patients with left ventricular failure, and that death rates are reduced [24]. These findings are an important indicator that renin-angiotensin inhibition is crucial to improving cardiac dysfunction. When the relationship of RAAS with myocardial infarction is examined, it has been determined that ACE2 RNA expression increases in the case of myocardial infarction [25]. In another study, it was shown that ACE2 expression increased in the case of myocardial injury induced by ischemia–reperfusion in rats and this increase attenuated myocardial damage [26].

2.2 Hypertension

It has been determined that the plasma renin level changes in the case of hypertension. Plasma renin levels are not proportional to blood pressure, and it has been reported that plasma renin levels are low in some patients, normal in others and high in others. One of the reasons for the change in the renin level is that it is primarily caused by ischemia that develops in the nephrons. In this case, renin levels released from ischemic nephrons increase at different levels, resulting in normal or high plasma renin levels. The renin released from ischemic nephrons passes into the circulation leading to the formation of AngII [17, 27]. As a result, hypertension occurs with increased vasoconstriction and sodium retention in nephrons. The reason why plasma renin level is normal in some hypertensive patients is that aldosterone is not synthesised in response to sodium restriction. Also, it has been stated that resistance to renin and AngII is formed in the vessels and therefore they can increase blood pressure even at low levels. Besides, independent of RAAS in circulating blood, it has been determined that Ang II production by serine protein kinase activity is independent of ACE activity in the heart, brain, adrenal cortex and blood vessels [28]. Also, AngII contributes to hypertension [29]. When looking at the relationship between salt intake and RAAS, it is seen that high salt intake suppresses RAAS, while low salt intake stimulates AngII release [30]. Studies have determined that smooth muscle cells are also critical in the regulation of AngII-mediated blood pressure. A study in mice found that 22α protein deficiency in smooth muscle reduces hypertension that can occur with AngII [31]. This is an indication that the RAAS system plays an important role in hypertension.

2.3 Atherosclerosis

AngII has been determined to induce endothelial dysfunction and increase oxidative stress in the endothelium by stimulating the production of reactive oxygen species (ROS) such as superoxide anions (O2−) derived from nicotinamide adenine dinucleotide phosphate oxidase (NADPH oxidase). This is especially the result of endothelial AT1R stimulation that interacts with the Nox5/Ca2 + calmodulin binding site, which will increase Ca + concentration in the endothelial cell [32, 33]. Nox5 is a member of the NADPH oxidase family and plays an important role in the development of atherosclerosis, inflammation, and oxidative stress [33, 34]. It also plays a role in the adhesion of mononuclear cells to the arterial endothelium and recruitment of mononuclear cells by stimulating the increase in CAM expression of TNF-α, which is released as a result of stimulation of AT1R with AngII, in combination with IL-6 [35]. One study reported that AngII induced monocyte chemotactic protein-derived protein expression (MCPIP1) via an AMPK/p38 MAPK-dependent pathway [36]. Increased MCPIP1 expression contributes to atherosclerotic plaque formation by triggering apoptosis in macrophages [37]. Another thing related to the formation of atherosclerosis is that AngII induces the expression of a multi-functional protein found in macrophages, endothelial cells, smooth muscle cells (SMCs), and epithelial cells called osteopontin. Osteopontin plays an important role in the development and development of atherosclerosis [38]. The cell membrane has a transmembrane glycoprotein called LOX. LOX acts as a receptor for oxidised LDL (oxLDL). It increases the expression of AngII LOX-1 gene. Binding of oxLDL to LOX-1 in the endothelium causes an increase in leukocyte adhesion molecules, activates apoptosis pathways, increases ROS and induces endothelial dysfunction. This situation contributes to the development of atherosclerosis. Also, oxLDL increases the formation of ACE, which induces the formation of AngII (**Figure 2**). This increases LOX-1 expression, which positively regulates the expression of AT1R, and contributes to a self-sustaining pro-atherogenic cycle [39]. Thus, it has been determined that ACE and ATR1 inhibitors prevent the development of atherosclerosis.

2.4 Vascular inflammation

RAAS plays an important role in shaping vascular inflammation. Vascular inflammation causes endothelial dysfunction. This dysfunction causes tissue damage. Endothelial dysfunction also results in the accumulation of inflammatory cells in the area. This situation triggers atherosclerosis. Also, studies have shown that

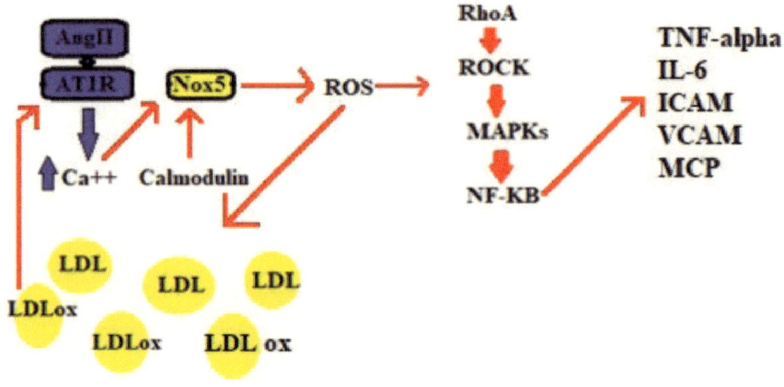

Figure 2.
Mechanism of AngII-mediated atherosclerosis formation. Involvement of Ang-II, ACE2, and Ang-1–7 in atherogenic pathways. The Ang-II binding into AT1R can activate Nox5 through a calcium/calmodulin-dependent pathway.

AngII-mediated inflammation and hypertension and atherosclerosis develop [40]. In another study, it was determined that AngII administration in human vascular smooth muscle cells increased NF-KB activation, thus increasing IL-6, MCP-1 and TNF-expression [41]. Again, although it is a vasoconstrictor, AngII was determined to induce endothelial damage by inhibiting endothelial cell regeneration. AngII has been reported to act as a second messenger to activate intracellular signalling pathways such as mitogen-activated protein kinase (MAPK) and protein kinase Akt/protein kinase B (Akt/PKB), pathways that mediate cell proliferation and apoptosis, and thus vascular dysfunction [42]. AngII is also stated to be a potent pro-oxidant. Ang II induces the production of superoxide anions and activates NADH/NADPH signalling [43]. AngII lowers nitric oxide (NO) levels and activates redox-sensitive genes, particularly cytokines and adhesion molecules [44]. Ang II is also a profibrotic factor. Chronic AngII administration in mice has been shown to cause an increase in blood pressure, infiltration of inflammatory cells into the myocardium and cardiac fibrosis [45]. Another factor that provides the proinflammatory and profibrinolytic effect of RAAS in vessels is aldosterone [46]. Aldosterone affects insulin resistance and the development of atherosclerosis. In vascular smooth muscle cells, aldosterone alters insulin signalling, increases insulin-like growth factor-1 expression.

2.5 Oxidative stress

Oxidative stress is defined as the disproportion between the presence of antioxidants and free radicals or prooxidants in a biological system. ROS and reactive nitrogen species (RNTs) are by-products of a variety of cellular processes, including aerobic metabolism [47–51]. These by-products cause damage to various tissues [52–73]. RAAS has a direct relationship with oxidative stress that may occur in the cardiovascular system. It has been determined that chronic administration of aldosterone, one of the components of RAAS, causes oxidative stress in the rat aorta [74]. AngII represents one of the major vasoactive peptides involved in the regulation and activation of NADPH oxidase. Ang II stimulates the activation of NADPH oxidase, increases the expression of NADPH oxidase subunits, and induces ROS formation in vascular smooth muscle cells, endothelial cells and fibroblasts. ACE2 shows an effect of reducing oxidative stress by inhibition of ROS synthesis by reducing AngII to Ang 1–7. Ang 1–7 therapy can have a curative effect on vascular disease models. It is reported that solutions that can increase Ang 1–7 levels may be beneficial to alleviate endothelial dysfunction [75]. This is supported by studies showing that overexpression of ACE2 leads to attenuating the effects of hypertension in animal models [76, 77]. It supports the argument that hypertension is a side effect directly related to oxidative stress, thus overexpression of ACE2 leads to a reduction of oxidative stress in a biological system [78].

3. The role of RAAS in renal diseases

3.1 Proteinuria

RAAS plays an important role in the pathogenesis of many kidney diseases characterised by proteinuria. In a study, it was stated that AngII induces the formation of proteinuria. It has also been determined that AngII stimulates the formation of TGF-1 in various kidney cells [79]. TGF-1 has been found to impair autoregulation by afferent arterioles [80]. Vasoconstriction occurs after increased arterial pressure in afferent arterioles. In case of impaired autoregulation in the presence of TGF-1, especially systemic hypertension occurs, an increase in transcapillary pressure occurs. Thus, AngII increases capillary filtration pressure by causing efferent vasoconstriction and TGF-1-mediated impaired afferent arteriole autoregulation. Also, AngII has been found to have a direct effect on the integrity of the filtration barrier. Again, AngII has been shown to reduce the synthesis of negatively charged proteoglycans and additionally suppress nephrin synthesis [81]. It has been observed that this situation causes apoptosis in podocytes. Vascular endothelial growth factor (VEGF) has been identified to be an important factor in increasing the permeability of the filtration barrier in the kidneys [82]. It has been determined to stimulate VEGF expression via the AngII, AT1 and AT 2 receptors. It is thought that the increase in VEGF expression via AT2 receptors may be mediated by an increase in hypoxia-inducible factor 1. Also, VEGF and TGF-1 mediate the AngII-mediated synthesis of the 3rd chain of collagen type IV, which is a component of the glomerular basement membrane [83, 84]. As a result, it is seen that AngII causes proteinuria by causing changes in hemodynamic and non-hemodynamic mechanisms. AngII stimulates albumin reabsorption in proximal tubule cells through AT2 receptor-mediated protein kinase B activation [85]. Albumin uptake induces a selection of proinflammatory and profibrogenic cytokines such as monocyte chemoattractant protein-1, IL-8, endothelin, and TGF-1 [86]. This situation stimulates the migration of cells into the interstitium. Ultimately it causes inflammation in the interstitial area.

3.2 Fibrosis

In a study, ECM proteins induce type I procollagen and mRNA encoding fibronectin in cultured mesangial cells of AngII, and also stimulates the synthesis of type I collagen types 1 and 3 in cultured proximal tubular cells [79]. It has been determined that the stimulatory effect of AngII on collagen expression is dependent on TGF-1 expression. As a result of the studies, it has been reported that AngII stimulates the proliferation of cultured renal fibroblasts and increases mRNA expression of TGF-β1, fibronectin and type I collagen. It has also been observed that renin increases TGF-1 expression by stimulating a particular receptor in cultured mesangial cells [87]. These findings suggest that increased renin as a result of ACE inhibitor therapy may directly contribute to renal fibrosis through increased TGF-1 despite AngII blockade. It was also determined that AngII increased connective tissue growth factor (CTGF) in kidney tissue. CTGF is a fibrinolytic mediator and is also stimulated by TGF-β. However, AngII also stimulates CTGF synthesis independently of TGF-β [88]. These findings suggest that increased renin as a result of ACE inhibitor therapy may directly contribute to renal fibrosis through increased TGF-1 despite AngII blockade. It was also determined that AngII increased connective tissue growth factor (CTGF) in kidney tissue. CTGF is a fibrinolytic mediator and is also stimulated by TGF-β. However, AngII also stimulates CTGF synthesis independently of TGF-β [89]. Studies have shown that more than one-third of local fibroblasts in renal interstitial fibrosis originate from tubular epithelial cells through a process called epithelial to mesenchymal transition (EMT). Again, AngII can be effective on EMT [90].

3.3 Inflammation

Studies have shown that AngII activates the proinflammatory transcription factor NF-KB via AT1 and AT2 [91]. It has also been stated that it can stimulate

NF-KB in AngIII and AngIV [86]. It has been determined that Rho-kinase plays a role in AngII mediated NF-KB activation. Also, AngII stimulates the transcription factor Ets. This factor regulates vascular inflammation by the transport of T cells and macrophages to the vascular wall. AngII has been reported to increase the level of Toll-like 4 receptors that bind LPS on mesangial cells. It has been observed that this receptor has an increasing effect on NF-KB activation [92]. The penetration of inflammatory cells into the glomerulus as well as the tubulointerstitium plays an important role in the progression of chronic kidney disease. Also, AngII induces the adhesion of circulating immune cells to capillaries by stimulating the increase of adhesion molecules such as vascular cellular adhesion molecule-1, intracellular adhesion molecule-1 and integrins. This situation shows the relationship of AngII with renal inflammation. It has also been determined that AngII has a stimulating effect on lymphocyte production [86, 93].

3.4 Chronic kidney disease (CKD)

Studies explaining the relation of RAAS with CKD were made in the 1980s and important data were obtained in these studies [94]. AngII has emerged as a central mediator of kidney damage because it can induce glomerular capillary hypertension that damages endothelial, glomerular epithelial cells, and mesangial cells [94, 95]. Also, AngII/aldosterone has non-haemodynamic effects that are important in the pathogenesis of CKD, such as inflammation, fibrosis, ROS production, and activation of pathways associated with endothelial dysfunction [94]. One of the most common causes of CKD is diabetic nephropathy. RAAS has an important role in diabetic nephropathy. Plasma renin activity is lower than normal in patients with diabetes [96]. However, intra-renal RAAS activity is high [97, 98]. This is an indication that diabetic nephropathy has one of the most important roles in the formation of CKD.

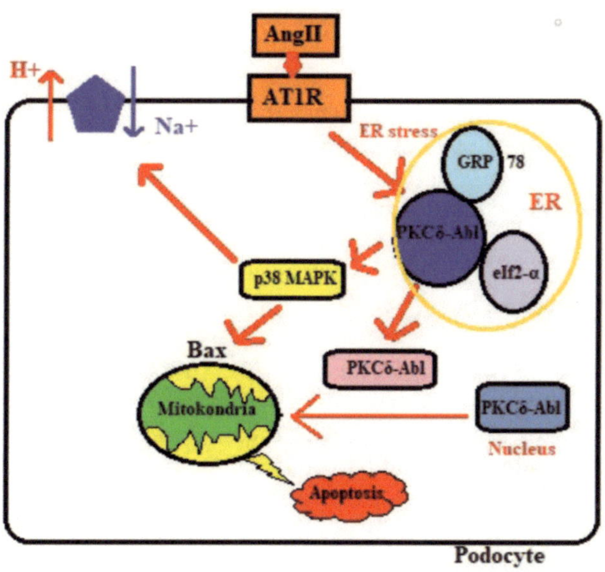

Figure 3.
Mechanism of AngII-mediated apoptosis formation in the podocyte. AT1R signalling induces ER stress through increased GRP 78 and p-elf2α expression and PKC-δ phosphorylation. p38 MAPK and PKC-δ activation lead to increased Bax expression and enhanced NHE1 activity, triggering cellular apoptosis.

3.5 Apoptosis

Studies show that the RAAS system is associated with renal hypertrophy and apoptosis. It has been determined that AngII, one of the components of RAAS, induces apoptosis in vivo and in vitro conditions [99]. It has been reported that AT1 and AT2 receptors are involved in these effects. Studies have reported that Ang II plays an important role in tubular cells and podocytes in (Endoplasmic reticulum) ER stress-induced renal apoptosis, especially in diabetic nephropathy [100]. It has been shown that Ang II can induce podocyte ER stress via the PERK-eIf2-α-ATF4 axis and the PI3-kinase pathway [101]. Another study found an AT1R-mediated increase in glomerular GRP 78 in rats chronically treated with AngII. These data support the relationship between the AngII/AT1R signal and ER stress on podocyte damage. In the same study, Ang II treatment was reported to induce p38 MAPK-dependent apoptosis in podocytes associated with Bax protein activation. In addition, Na+/H+ exchanger isoform 1 (NHE1) activity increases. As a result, it triggers cellular apoptosis (**Figure 3**), [102].

4. The role of RAAS in lung diseases

4.1 Acute lung injury and pneumonia

As a result of RAAS activation, inflammation [103] and vascular permeability increase [104] due to Ang II stimulation of AT1 receptor and thus severe acute lung damage occurs [105, 106]. In mice, administration of losartan prevents acute lung injury caused by Ang II and decreases AT1R expression [107, 108]. Pneumonia is associated with RAAS, especially in influenza-induced types of pneumonia RAAS system plays a very important role. In patients with pneumonia, the use of RAAS inhibitors reduces the mortality rate and the likelihood of intubation [109]. As with other viral types of pneumonia, children infected with the Respiratory syncytial virus (RSV) tend to have higher Ang II levels than healthy children [110]. The benefit of recombinant ACE2 treatment on RSV infection has been demonstrated in a preclinical mouse model in animal experiments [111]. H7N9 and H5N1 influenza reduce the level of ACE2, increase the level of Ang II, and thus cause lung damage via the AT1 receptor [112]. In H5N1 and H7N9 mouse models, treatment with losartan results in a decrease in IL-6 level and lung oedema, thus preventing lung damage [113]. It was concluded that losartan prevents lung damage by inhibiting RAAS activity.

4.2 SARS-CoV viral infection

The Spike protein [S protein] on the SARS-CoV Virus surface attaches to the ACE2 receptor and enters the body in this way. Moreover, ACE2 improves the efficiency of SARS-CoV replication [114]. Transmembrane protease serine 2 (TMPRSS2) can degrade ACE2 and S protein for membrane fusion and the entry of SARS-CoV into cells. Therefore, the concentration of ACE2 in the membrane decreases, but the number of cells infected with SARS-CoV with cessation increases [115]. Ang-II level increases in lung tissue of mice infected with SARS-CoV. Also, the use of angiotensin receptor blockers in these animals significantly reduces pulmonary oedema. This indicates that lung failure caused by SARS-CoV is caused by an increase in Ang-II level and overactivation of the AT1 receptor [116]. Increased ACE level and decreased ACE2 levels in SARS patients cause increased Ang II level and AT1 receptor expression, which accelerates lung damage and can lead to death [117]. Also, tumour necrosis factor-α (TNF-α), interleukin-6 (IL-6), interleukin-8 (IL-8), caspase 3 (CASP3), caspase 9 (CASP9) and fibroblast growth factor-7 (FGF-7) increase in the lung tissue of these patients [118].

4.3 SARS-CoV-2 viral infection

SARS-CoV-2 (Covid-19) Similar to SARS-CoV, the S protein uses the ACE2 cellular membrane for input and uses TMPRSS2 for S protein preparation to facilitate the fusion of viral and cellular membranes [119–121]. Compared to other coronaviruses, the affinity of S protein to ACE2 is higher in SARS-CoV and SARS-CoV-2. Looking at the distribution of ACE2 receptors in the body, it is found on the endothelial cells and smooth muscle cells of organs and tissues, including the oral and nasal mucosa, lung, small intestine, kidney, heart and blood vessels. The widespread distribution of ACE2 receptors in the body is an indicator of multi-organ failure in COVID-19 patients [122–124]. SARS-CoV-2 infection causes RAAS disorders and systemic inflammatory response. The plasma Ang II level of COVID-19 patients is significantly higher than that of healthy individuals. This condition is linearly related to viral load and lung injury [125]. A clinical study has shown that cytokine storm syndrome (CSS) occurs in patients with COVID-19 and severe pneumonia. Also, it showed that some cases can progress rapidly to Acute respiratory distress syndrome (ARDS) and even to multiple organ failure [126]. Inflammatory cytokines and chemokines are synthesised in Covid-19 patients, including IL-6, IL-2, IL-1β, IL-8, IL-17, IFN-γ, TNF-α and monocyte chemoattractant protein-1 (MCP-1) (**Figure 4**). Among them, however, IL-6 in particular plays a key role in triggering the inflammatory response, increasing the mortality rate in patients [125]. In Covid-19 infection, after the virus binds to ACE2 on the cell surface, Ang II cannot convert to Ang1–7, and thus more and more binding occurs to AT1 receptors. This situation causes an imbalance in the ACE/ Ang II/AT1R axis. As a result, the pulmonary endothelium and epithelial cells are damaged by stimulating inflammatory signalling pathways, resulting in an increase in the permeability of pulmonary capillaries [127].

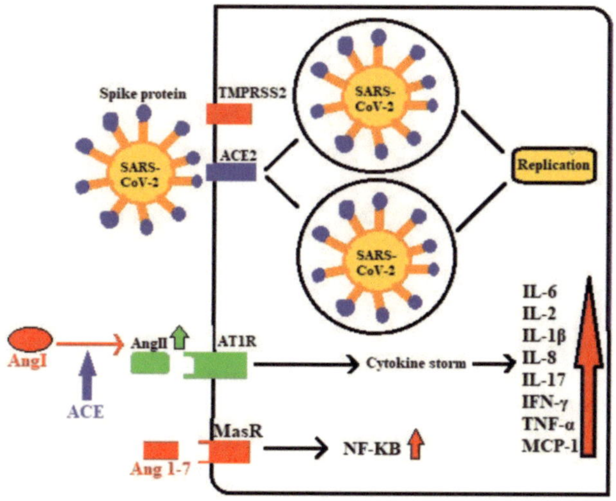

Figure 4.
Effects of the renin-angiotensin system during SARS-CoV-2 infection.

5. The role of RAAS in some neurological disorders

Brain RAAS irregularity may contribute to neurodegeneration due to neuroinflammation, oxidative stress and pathophysiological changes due to ageing. Several studies have reported that irregular RAAS plays a key role in numerous degenerative diseases of the brain, including Alzheimer's, Parkinson's disease, Huntington's disease, dementia, amyotrophic lateral sclerosis, Multiple sclerosis, Traumatic brain injury, and Stroke [128–130].

5.1 Alzheimer's disease

Alzheimer's disease (AD) is a progressive neurodegenerative disease characterised by impaired daily functions and behaviour, especially memory [131]. The most important change in AD neuropathology is Aβ-centred senile amyloid plaques formed in the hippocampus, amygdala and cortex. Neurovascular disorders and chronic neurodegeneration occur in the surrounding brain tissues and vessels as a result of the toxic effects of these plaques [132]. Besides these plaque formations; Neurofibrillary tangles, oxidative stress in cell membranes and organelles, inflammation, gliosis, excitotoxicity due to excessive intracellular Ca + 2 increase and neuron death by many mechanisms that trigger each other such as disruption in membrane cation channels are encountered [133, 134]. The amyloid-beta (Aβ) peptide triggers O2 radical production in endothelial cells and induces oxidative and peroxidative reactions, causing cell death. As an example of these reactions; the oxidative reaction catalysed by the combination of amyloid plaques with heavy metal ions and lipid membrane peroxidation by various mechanisms can be given. It has been observed that the increased ROS activity via Aβ in tissue taken from the hippocampus caused synaptic disruption and cell death as a result of increased Ca + 2 increase with N-methyl-D-aspartate (NMDA) channel activation. Besides, mitochondria dysfunction is an important point in AD pathology. In biopsy studies, it was found that mitochondria shrank and protein and DNA dispersed into the cytoplasm [135, 136].

One of the brain RAAS products, the Ang- (1–7) peptide is a Mas receptor [MASR] agonist [137]. MASRs are abundant in memory-related areas of the brain and accelerate hippocampal long-term potentiation (LTP) together with Ang-(1–7). Also, it is known that the neuroinflammatory effects of Ang II, another RAAS product, contribute to cognitive disorders. Reversing the biological effects of Ang II with the anti-inflammatory, anti-fibrotic, vasodilator and anti-proliferative biological effects of Ang- (1–7); supports memory and learning [138]. In brain tissue studies in AD, it has been shown that the expression and activity of ACE, the metabolic enzyme of Ang-II, changes significantly in certain regions of the brain, including the frontal cortex and hippocampus. It has been reported that when centrally acting ACE inhibitors are used, they have reduced cognitive decline and have memory-enhancing effects [139, 140]. ACE2 activity decreases in AD pathology [141]. Ang- (1–7) improves memory functions without affecting hippocampal or cortical amyloid peptide storage [142].

Ang II causes oxidative stress through the AT1 receptor [143] and increases superoxide. Thus, it causes neuroinflammation and vascular diseases [144]. As a result, it causes Aβ accumulation due to AD. However, the AT2 receptor signal produces beneficial effect including learning and memory. Angiotensin receptor blockers (ARBs) inhibit AT1R signalling, which shifts the effect of Ang-II towards the beneficial path (AT2R signal) (**Figure 5**) [144].

ACE inhibitors have a protective effect against AD. It shows this effect by suppressing brain-derived neurotrophic factor reduction and TNF-α release.

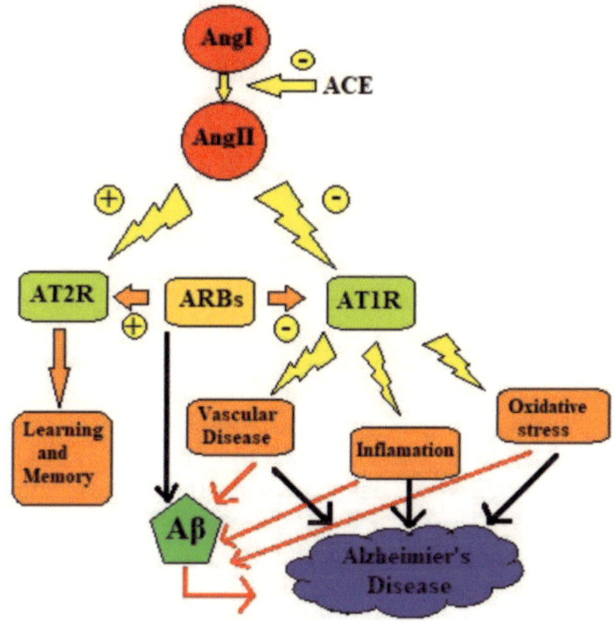

Figure 5.
Effect of AngII on the nervous system. Amyloid plaque (Aβ), angiotensin II (AngII), angiotensin I (AngI), angiotensin-converting enzyme inhibitors (ACEIs), angiotensin receptor blockers (ARBs), angiotensin (AT), AT2 receptor (AT2R), AT1 receptor (ATR1R).

ACE inhibitors also improve oxidative-nitrosative stress and nitrotyrosine production, which reduces amyloidogenesis and subsequent Aβ accumulation [145, 146]. On the other hand, ACE inhibitor (Captopril) and Angiotensin receptor blockers (Telmisartan, Candesartan, Losartan) ameliorate oxidative stress [147–151]. Telmisartan normalises the decreased thioredoxin (TRX) system in addition to attenuating the expression of the protein (TXNIP) that interacts with thioredoxin. Thus, it reduces the formation of endogenous ROS [149]. Similarly, telmisartan reduces improved glycation end products and 4-hydroxynonenal, which are markers of oxidative stress and are associated with Neurodegeneration [150]. Candesartan lowers the level of free radicals in the brain by decreasing malondialdehyde and increasing glutathione levels [151].

5.2 Parkinson's disease

Ang II levels are high in the striatum and substantia nigra of Parkinson's disease (PD) patients. Ang II and AT1R trigger apoptosis by activating autophagy in a dopaminergic neuronal cell. These findings suggest that Ang II plays a role in the pathogenesis of PD [152]. In animal models of PD, it has been found that the signalling of AT2Rs is decreased with the loss of function in dopaminergic neurons [153]. Also, ACE and ACE2 were detected in the cerebrospinal fluid of PD patients. ACE levels are decreased in the cerebrospinal fluid of PD patients [154].

5.3 Multiple sclerosis

Multiple sclerosis (MS) is defined as an autoimmune neurodegenerative disease that typically occurs in the third or fourth decade of life [155]. Although the aetiology of the disease is not fully known, both environmental and genetic factors are thought to play an important role in the development of MS [156]. Blocking angiotensin II production by ACE inhibitors and inhibition of angiotensin II signalling by AT1 receptor blockers suppresses T-helper 17 (Th17) cells [157]. Th17 cells play an important role in the development and relapse of MS [158]. In a study, ACE activity in the blood serum of MS patients was reported to be higher than in healthy controls [159]. In another study, ACE and ACE2 levels were found to be reduced in the cerebrospinal fluid of MS patients [160].

6. Conclusion

As understood, the renin-angiotensin-aldosterone system plays a very important role in regulating the fluid-electrolyte balance and blood pressure in the body. RAAS has receptors in many organs and tissues and can show various effects here. RAAS can be affected by various diseases affecting the cardiovascular, renal, nervous and respiratory systems and plays a major role in the formation of damage that may occur in these systems. Drugs that can affect the components or receptors of RAAS can prevent damage that may occur. The presented study shows the importance of the role of this system in the mentioned diseases. Understanding the role of this system in the mentioned diseases is of great importance in the development of new treatment protocols and new therapeutic agents.

Author details

Volkan Gelen[1*], Abdulsamed Kükürt[2] and Emin Şengül[3]

1 Department of Physiology, Faculty of Veterinary Medicine, Kafkas University, Kars, Turkey

2 Department of Biochemistry, Faculty of Veterinary Medicine, Kafkas University, Kars, Turkey

3 Department of Physiology, Faculty of Veterinary Medicine, Atatürk University, Erzurum, Turkey

*Address all correspondence to: gelen_volkan@hotmail.com

References

[1] Tigerstedt R, Bergman PQ. Niere und Kreislauf 1. Skand Arch Physiol 1898;8:223-71. https://doi.org/10.1111/j.1748-1716.1898.tb00272.x.

[2] Peach MJ. Renin-angiotensin system: biochemistry and mechanisms of action. Physiol Rev 1977;57:313-70. https://doi.org/10.1152/physrev.1977.57.2.313.

[3] Leung PS, Chappell MC. A local pancreatic renin-angiotensin system: endocrine and exocrine roles. Int J Biochem Cell Biol 2003;35:838-46. https://doi.org/10.1016/S1357-2725(02)00179-6.

[4] Skøtt O, Jensen BL. Cellular and intrarenal control of renin secretion. Clin Sci 1993;84:1-10. https://doi.org/10.1042/cs0840001.

[5] Grote K, Drexler H, Schieffer B. Renin-angiotensin system and atherosclerosis. Nephrol Dial Transplant 2004. https://doi.org/10.1093/ndt/gfh030.

[6] Naftilan AJ, Zuo WM, Inglefinger J, Ryan TJ, Pratt RE, Dzau VJ. Localization and differential regulation of angiotensinogen mRNA expression in the vessel wall. J Clin Invest 1991;87:1300-11. https://doi.org/10.1172/JCI115133.

[7] Csikós T, Chung O, Unger T. Receptors and their classification: focus on angiotensin II and the AT2 receptor. J Hum Hypertens 1998;12:311-8. https://doi.org/10.1038/sj.jhh.1000639.

[8] Erdös EG. Angiotensin I converting enzyme and the changes in our concepts through the years. Lewis K. Dahl memorial lecture. Hypertension 1990;16:363-70. https://doi.org/10.1161/01.HYP.16.4.363.

[9] Danilczyk U, Eriksson U, Crackower MA, Penninger JM. A story of two ACEs. J Mol Med 2003. https://doi.org/10.1007/s00109-003-0419-x.

[10] Guang C, Phillips RD, Jiang B, Milani F. Three key proteases – angiotensin-I-converting enzyme (ACE), ACE2 and renin – within and beyond the renin-angiotensin system. Arch Cardiovasc Dis 2012;105:373-85. https://doi.org/10.1016/j.acvd.2012.02.010.

[11] Wei L, Clauser E, Alhenc-Gelas F, Corvol P. The two homologous domains of human angiotensin I-converting enzyme interact differently with competitive inhibitors. J Biol Chem 1992. https://doi.org/10.1016/S0021-9258(18)42224-7.

[12] Donoghue M, Hsieh F, Baronas E, Godbout K, Gosselin M, Stagliano N, et al. A novel angiotensin-converting enzyme-related carboxypeptidase (ACE2) converts angiotensin I to angiotensin 1-9. Circ Res 2000. https://doi.org/10.1161/01.res.87.5.e1.

[13] Tipnis SR, Hooper NM, Hyde R, Karran E, Christie G, Turner AJ. A Human Homolog of Angiotensin-converting Enzyme. J Biol Chem 2000;275:33238-43. https://doi.org/10.1074/jbc.M002615200.

[14] Appel GB, Appel AS. Angiotensin II receptor antagonists: Role in hypertension, cardiovascular disease, and renoprotection. Prog Cardiovasc Dis 2004;47:105-15. https://doi.org/10.1016/j.pcad.2004.04.005.

[15] Reid IA. The renin-angiotensin system: physiology, pathophysiology, and pharmacology. Am J Physiol 1998. https://doi.org/10.1152/advances.1998.275.6.s236.

[16] Bader M, Ganten D. Update on tissue renin-angiotensin systems. J Mol

Med 2008. https://doi.org/10.1007/s00109-008-0336-0.

[17] Goodfriend TL, Elliott ME, Catt KJ. Angiotensin Receptors and Their Antagonists. N Engl J Med 1996;334:1649-55. https://doi.org/10.1056/NEJM199606203342507.

[18] Tham YK, Bernardo BC, Ooi JYY, Weeks KL, McMullen JR. Pathophysiology of cardiac hypertrophy and heart failure: signaling pathways and novel therapeutic targets. Arch Toxicol 2015;89:1401-38. https://doi.org/10.1007/s00204-015-1477-x.

[19] Urata H, Healy B, Stewart RW, Bumpus FM, Husain A. Angiotensin II-forming pathways in normal and failing human hearts. Circ Res 1990;66:883-90. https://doi.org/10.1161/01.RES.66.4.883.

[20] Sadoshima J, Izumo S. Molecular characterization of angiotensin II--induced hypertrophy of cardiac myocytes and hyperplasia of cardiac fibroblasts. Critical role of the AT1 receptor subtype. Circ Res 1993;73:413-23. https://doi.org/10.1161/01.RES.73.3.413.

[21] Clavell A, Stingo A, Margulies K, Lerman A, Underwood D, Burnett JC. Physiological significance of endothelin: Its role in congestive heart failure. Circulation 1993;87:V45-50.

[22] Holubarsch C, Hasenfuss G, Schmidt-Schweda S, Knorr A, Pieske B, Ruf T, et al. Angiotensin I and II exert inotropic effects in atrial but not in ventricular human myocardium. An in vitro study under physiological experimental conditions. Circulation 1993;88:1228-37. https://doi.org/10.1161/01.CIR.88.3.1228.

[23] Lopez JJ, Lorell BH, Ingelfinger JR, Weinberg EO, Schunkert H, Diamant D, et al. Distribution and function of cardiac angiotensin AT1- and AT2-receptor subtypes in hypertrophied rat hearts. Am J Physiol Circ Physiol 1994;267:H844-52. https://doi.org/10.1152/ajpheart.1994.267.2.H844.

[24] Burrell LM, Risvanis J, Kubota E, Dean RG, MacDonald PS, Lu S, et al. Myocardial infarction increases ACE2 expression in rat and humans. Eur Heart J 2005;26:369-75. https://doi.org/10.1093/eurheartj/ehi114.

[25] Der Sarkissian S, Grobe JL, Yuan L, Narielwala DR, Walter GA, Katovich MJ, et al. Cardiac Overexpression of Angiotensin Converting Enzyme 2 Protects the Heart From Ischemia-Induced Pathophysiology. Hypertension 2008;51:712-8. https://doi.org/10.1161/HYPERTENSIONAHA.107.100693.

[26] Laragh JH. Lewis K. Dahl Memorial Lecture. The renin system and four lines fo hypertension research. Nephron heterogeneity, the calcium connection, the prorenin vasodilator limb, and plasma renin and heart attack. Hypertension 1992;20:267-79. https://doi.org/10.1161/01.HYP.20.3.267.

[27] Williams GH, Fisher NDL, Hunt SC, Jeunemaitre X, Hopkins PN, Hollenberg NK. Effects of gender and genotype on the phenotypic expression of nonmodulating essential hypertension. Kidney Int 2000;57:1404-7. https://doi.org/10.1046/j.1523-1755.2000.00982.x.

[28] Batchu SN, Hughson A, Wadosky KM, Morrell CN, Fowell DJ, Korshunov VA. Role of Axl in T-Lymphocyte Survival in Salt-Dependent Hypertension. Arterioscler Thromb Vasc Biol 2016;36:1638-46. https://doi.org/10.1161/ATVBAHA.116.307848.

[29] Gao S, Cui X, Wang X, Burg MB, Dmitrieva NI. Cross-Sectional Positive Association of Serum Lipids and Blood Pressure With Serum Sodium Within the Normal Reference Range of 135-145

mmol/L. Arterioscler Thromb Vasc Biol 2017;37:598-606. https://doi.org/10.1161/ATVBAHA.116.308413.

[30] Carillo BA, Beutel A, Mirandola DA, Vidonho AF, Furukawa LNS, Casarini D, et al. Differential sympathetic and angiotensinergic responses in rats submitted to low- or high-salt diet. Regul Pept 2007;140:5-11. https://doi.org/10.1016/j.regpep.2006.11.007.

[31] Ziegler T, Abdel Rahman F, Jurisch V, Kupatt C. Atherosclerosis and the Capillary Network; Pathophysiology and Potential Therapeutic Strategies. Cells 2019;9:50. https://doi.org/10.3390/cells9010050.

[32] Montezano AC, Burger D, Paravicini TM, Chignalia AZ, Yusuf H, Almasri M, et al. Nicotinamide Adenine Dinucleotide Phosphate Reduced Oxidase 5 (Nox5) Regulation by Angiotensin II and Endothelin-1 Is Mediated via Calcium/Calmodulin-Dependent, Rac-1-Independent Pathways in Human Endothelial Cells. Circ Res 2010;106:1363-73. https://doi.org/10.1161/CIRCRESAHA.109.216036.

[33] Piqueras L, Sanz M-J. Angiotensin II and leukocyte trafficking: New insights for an old vascular mediator. Role of redox-signaling pathways. Free Radic Biol Med 2020;157:38-54. https://doi.org/10.1016/j.freeradbiomed.2020.02.002.

[34] Touyz RM, Anagnostopoulou A, Rios F, Montezano AC, Camargo LL. NOX5: Molecular biology and pathophysiology. Exp Physiol 2019;104:605-16. https://doi.org/10.1113/EP086204.

[35] Shu S, Zhang Y, Li W, Wang L, Wu Y, Yuan Z, et al. The role of monocyte chemotactic protein-induced protein 1 (MCPIP1) in angiotensin II-induced macrophage apoptosis and vulnerable plaque formation. Biochem Biophys Res Commun 2019;515:378-85. https://doi.org/10.1016/j.bbrc.2019.05.145.

[36] Silva GM, França-Falcão MS, Calzerra NTM, Luz MS, Gadelha DDA, Balarini CM, et al. Role of Renin-Angiotensin System Components in Atherosclerosis: Focus on Ang-II, ACE2, and Ang-1-7. Front Physiol 2020;11. https://doi.org/10.3389/fphys.2020.01067.

[37] Ding Y, Chen J, Cui G, Wei Y, Lu C, Wang L, et al. Pathophysiological role of osteopontin and angiotensin II in atherosclerosis. Biochem Biophys Res Commun 2016;471:5-9. https://doi.org/10.1016/j.bbrc.2016.01.142.

[38] Lubrano V, Balzan S. Roles of LOX-1 in microvascular dysfunction. Microvasc Res 2016;105:132-40. https://doi.org/10.1016/j.mvr.2016.02.006.

[39] Kattoor AJ, Kanuri SH, Mehta JL. Role of Ox-LDL and LOX-1 in Atherogenesis. Curr Med Chem 2019;26:1693-700. https://doi.org/10.2174/0929867325666180508100950.

[40] Kranzhöfer R, Schmidt J, Pfeiffer CAH, Hagl S, Libby P, Kübler W. Angiotensin Induces Inflammatory Activation of Human Vascular Smooth Muscle Cells. Arterioscler Thromb Vasc Biol 1999;19:1623-9. https://doi.org/10.1161/01.ATV.19.7.1623.

[41] Becher UM, Endtmann C, Tiyerili V, Nickenig G, Werner N. Endothelial Damage and Regeneration: The Role of the Renin-Angiotensin-Aldosterone System. Curr Hypertens Rep 2011;13:86-92. https://doi.org/10.1007/s11906-010-0171-x.

[42] Rajagopalan S, Kurz S, Münzel T, Tarpey M, Freeman BA, Griendling KK, et al. Angiotensin II-mediated hypertension in the rat increases vascular superoxide production via membrane NADH/NADPH oxidase

activation. Contribution to alterations of vasomotor tone. J Clin Invest 1996;97:1916-23. https://doi.org/10.1172/JCI118623.

[43] Wassmann S, Nickenig G. Pathophysiological regulation of the AT1-receptor and implications for vascular disease. J Hypertens 2006;24:S15-21. https://doi.org/10.1097/01.hjh.0000220402.53869.72.

[44] Qi G, Jia L, Li Y, Bian Y, Cheng J, Li H, et al. Angiotensin II Infusion–Induced Inflammation, Monocytic Fibroblast Precursor Infiltration, and Cardiac Fibrosis are Pressure Dependent. Cardiovasc Toxicol 2011;11:157-67. https://doi.org/10.1007/s12012-011-9109-z.

[45] Martinez FA. Aldosterone Inhibition and Cardiovascular Protection: More Important Than it Once Appeared. Cardiovasc Drugs Ther 2010;24:345-50. https://doi.org/10.1007/s10557-010-6256-6.

[46] Cascella T, Radhakrishnan Y, Maile LA, Busby WH, Gollahon K, Colao A, et al. Aldosterone Enhances IGF-I-Mediated Signaling and Biological Function in Vascular Smooth Muscle Cells. Endocrinology 2010;151:5851-64. https://doi.org/10.1210/en.2010-0350.

[47] Kükürt A, Kuru M, Faruk Başer Ö, Karapehlivan M. Kisspeptin: Role in Female Infertility. In: Marsh C, editor. Sex Horm. [Working Title], IntechOpen; 2020. https://doi.org/10.5772/intechopen.94925.

[48] Kükürt A, Kuru M, Karapehlivan M. Nitrik Oksit, Nitrik Oksit Sentaz ve Dişi Üreme Sistemindeki Rolleri. In: Evereklioğlu C, editor. Sağlık Bilim. Alanında Akad. Çalışmalar - II, Gece Kitaplığı; 2020, p. 113-23.

[49] Kara A, Gedikli S, Sengul E, Gelen V, Ozkanlar S. Oxidative Stress and Autophagy. Free Radicals Dis., InTech; 2016. https://doi.org/10.5772/64569.

[50] Kükürt A. Doğal bir antioksidan olarak propolis tedavisinin koruyucu etkileri. In: Evereklioğlu C, editor. Sağlık Bilim. Teor. ve Araştırmalar II, Gece Kitaplığı; 2020, p. 501-15.

[51] Kuru M, Kükürt A, Oral H, Öğün M. Clinical Use of Progesterone and Its Relation to Oxidative Stress in Ruminants. In: Drevenšek G, editor. Sex Horm. Neurodegener. Process. Dis., InTech; 2018, p. 303-27. https://doi.org/10.5772/intechopen.73311.

[52] Gelen V, Şengül E, Gedikli S, Atila G, Uslu H, Makav M. The protective effect of rutin and quercetin on 5-FU-induced hepatotoxicity in rats. Asian Pac J Trop Biomed 2017;7:647-53. https://doi.org/10.1016/j.apjtb.2017.06.013.

[53] Gelen V, Şengül E, Yıldırım S, Atila G. The protective effects of naringin against 5-fluorouracil-induced hepatotoxicity and nephrotoxicity in rats. Iran J Basic Med Sci 2018;21:404-10. https://doi.org/10.22038/ijbms.2018.27510.6714.

[54] Deveci HA, Karapehlivan M, Kaya İ, Kükürt A, Alpay M. Protective role of caffeic acid phenethyl ester against to chlorpyrifos-ethyl acute poisoning. Ankara Üniversitesi Vet Fakültesi Derg 2015;62:255-60. https://doi.org/10.1501/Vetfak_0000002689.

[55] Kaya İ, Kaya MM, Kükürt A, Özcan A, Karaman M, Deveci HA, et al. Effect of Ellagic Acid on Some Oxidative Stress Parameters and Cyclooxygenase-2 Reactivity in Mice with Experimental Gastric Injury. Japanese J Gastroenterol Hepatol 2019;2:1-9.

[56] Kaya İ, Deveci HA, Karapehlivan M, Kükürt A. Investigation of oxidative stress index in pyridine and ellagic acid

treated mice. Eurasian J Vet Sci 2015;31:148-148. https://doi.org/10.15312/EurasianJVetSci.2015310971.

[57] Şengül E, Gelen V, Gedikli S, Özkanlar S, Gür C, Çelebi F, et al. The protective effect of quercetin on cyclophosphamide-Induced lung toxicity in rats. Biomed Pharmacother 2017;92:303-7. https://doi.org/10.1016/j.biopha.2017.05.047.

[58] Gelen V, Şengül E, Gedikli S, Gür C, Özkanlar S. Therapeutic effect of quercetin on renal function and tissue damage in the obesity induced rats. Biomed Pharmacother 2017;89:524-8. https://doi.org/10.1016/j.biopha.2017.02.057.

[59] Sengul E, Gelen V, Yildirim S, Tekin S, Dag Y. The Effects of Selenium in Acrylamide-Induced Nephrotoxicity in Rats: Roles of Oxidative Stress, Inflammation, Apoptosis, and DNA Damage. Biol Trace Elem Res 2021;199:173-84. https://doi.org/10.1007/s12011-020-02111-0.

[60] Gelen V, Şengül E. Antioxidant, anti-inflammatory and antiapoptotic effects of naringin on cardiac damage induced by cisplatin. Indian J Tradit Knowl 2020;19:459-65.

[61] Karamese M, Guvendi B, Karamese SA, Cinar I, Can S, Erol HS, et al. The protective effects of epigallocatechin gallate on lipopolysa ccharide-induced hepatotoxicity: An in vitro study on Hep3B cells. Iran J Basic Med Sci 2016;19:483-9. https://doi.org/10.22038/ijbms.2016.6932.

[62] Gedikli S, Gelen V, Sengul E, Ozkanlar S, Gur C, Agırbas O, et al. Therapeutic Effects of Melatonin On Liver And Kidney Damages In Intensive Exercise Model of Rats. Endocrine, Metab Immune Disord Targets 2015;15:308-14. https://doi.org/10.2174/1871530315666150827103043.

[63] Sengul E, Gelen V, Gedikli S. Cardioprotective Activities of Quercetin and Rutin in Sprague Dawley Rats Treated with 5-Fluorouracil. J Anim Plant Sci 2020;31:423-31. https://doi.org/10.36899/JAPS.2021.2.0231.

[64] Ogun M, Ozcan A, Karaman M, Merhan O, Ozen H, Kukurt A, et al. Oleuropein ameliorates arsenic induced oxidative stress in mice. J Trace Elem Med Biol 2016;36:1-6. https://doi.org/10.1016/j.jtemb.2016.03.006.

[65] Kükürt A, Gelen V, Başer ÖF, Deveci HA, Karapehlivan M. Thiols: Role in Oxidative Stress-Related Disorders.IntechOpen, 2021, https://doi.org/10.5772/intechopen.96682

[66] Gelen V, Şengül E, Çınar DA. The effects of rutin and quercetin on ECG parameters in 5-FU-induced cardiotoxicity rat model. World Journal of Advanced Research and Reviews 2021:09:253-257. https://doi.org/10.30574/wjarr.2021.9.3.0104

[67] Karamese M, Aydin H, Gelen V, Sengul E, Karamese SA. The anti-inflammatory, anti-oxidant and protective effects of probiotic mixture on organ toxicity in a rat model. Future Microbiol. 2020:15:401-12. https://doi.org/10.2217/fmb-2020-0005

[68] Sengul E & Gelen V, Protective effects of naringin in indomethacin-induced gastric ulcer in rats. GSC Biological and Pharmaceutical Sciences 2019:8:6-14. https://doi.org/10.30574/gscbps.2019.8.2.0132

[69] Gelen V, Sengul E, Yıldırım S, Celebi F, Cınar A. Effects of rutin on bladder contractility and histopathology in cyclophosphamide-induced hemorrhagic cystitis in rats. Ataturk University J Vet Sci 2018:13:337-346. https://dergipark.org.tr/.../607156

[70] Gelen V, Gelen S.U, Celebi F, Cinar A, Yildirim S, Eser G. The

protective effect of Lactobacillus rhamnosus, Lactobacillus fermentum and lactobacillus brevis against cisplatin-induced hepatic damage in rats. Fresenius Environ. Bull. 2019: 28:7583-7592. https://doi.org/10.30574/gscbps.2019.8.2.0132

[71] Uslu GA, Gelen V, Uslu H, Özen H. Effects of cinnamomum cassia extract on oxidative stress, immunreactivity of iNOS and impaired thoracic aortic reactivity induced by type II diabetes in rats. Brazilian J. Pharm. Sci 2018:54:1-9. https://doi.org/10.1590/s2175-97902018000317785

[72] Karamese M, Aydin H, Sengul E, et al. The Immunostimulatory Efect of Lactic Acid Bacteria in a Rat Model. Iranian journal of immunology: IJI 2016:13:220-228. https://doi.org/10.1590/s2175-97902018000317785

[73] Gedikli S, Ozkanlar S, Gur C, Sengul E, Gelen V. Preventive effects of quercetin on liver damages in highfat diet-induced obesity. Journal of Histology & Histopathology 2017: 4:7. http://dx.doi.org/10.7243/2055-091X-4-7

[74] Sherajee SJ, Fujita Y, Rafiq K, Nakano D, Mori H, Masaki T, et al. Aldosterone Induces Vascular Insulin Resistance by Increasing Insulin-Like Growth Factor-1 Receptor and Hybrid Receptor. Arterioscler Thromb Vasc Biol 2012;32:257-63. https://doi.org/10.1161/ATVBAHA.111.240697.

[75] Başer ÖF, Kükürt A, Karapehlivan M. Oksidatif stresin azaltılmasında anjiyotensin dönüştürücü enzimin rolü. In: Evereklioğlu C, editor. Sağlık Bilim. Teor. ve Araştırmalar II, Gece Kitaplığı; 2020, p. 243-53.

[76] Wysocki J, Ye M, Rodriguez E, González-Pacheco FR, Barrios C, Evora K, et al. Targeting the degradation of angiotensin II with recombinant angiotensin-converting enzyme 2: Prevention of angiotensin II-dependent hypertension. Hypertension 2010. https://doi.org/10.1161/HYPERTENSIONAHA.109.138420.

[77] Wysocki J, Ortiz-Melo DI, Mattocks NK, Xu K, Prescott J, Evora K, et al. ACE2 deficiency increases NADPH-mediated oxidative stress in the kidney. Physiol Rep 2014. https://doi.org/10.1002/phy2.264.

[78] Birben E, Sahiner UM, Sackesen C, Erzurum S, Kalayci O. Oxidative Stress and Antioxidant Defense. World Allergy Organ J 2012;5:9-19. https://doi.org/10.1097/WOX.0b013e3182439613.

[79] Wolf G. Link between Angiotensin II and TGF-β in the Kidney. Miner Electrolyte Metab 1998;24:174-80. https://doi.org/10.1159/000057367.

[80] Sharma K, Cook A, Smith M, Valancius C, Inscho EW. TGF-β impairs renal autoregulation via generation of ROS. Am J Physiol Physiol 2005;288:F1069-77. https://doi.org/10.1152/ajprenal.00345.2004.

[81] Brinkkoetter P-T. Angiotensin II Type 1-Receptor Mediated Changes in Heparan Sulfate Proteoglycans in Human SV40 Transformed Podocytes. J Am Soc Nephrol 2004;15:33-40. https://doi.org/10.1097/01.ASN.0000102476.50041.44.

[82] Wolf G, Chen S, Ziyadeh FN. From the Periphery of the Glomerular Capillary Wall Toward the Center of Disease: Podocyte Injury Comes of Age in Diabetic Nephropathy. Diabetes 2005;54:1626-34. https://doi.org/10.2337/diabetes.54.6.1626.

[83] Wolf G, Schroeder R, Stahl RAK. Angiotensin II Induces Hypoxia-Inducible Factor-1α in PC 12 Cells through a Posttranscriptional Mechanism: Role of AT_2 Receptors. Am J Nephrol 2004;24:415-21. https://doi.org/10.1159/000080086.

[84] Chen S, Lee JS, Iglesias-de la Cruz MC, Wang A, Izquierdo-Lahuerta A, Gandhi NK, et al. Angiotensin II stimulates α3(IV) collagen production in mouse podocytes via TGF-β and VEGF signalling: implications for diabetic glomerulopathy. Nephrol Dial Transplant 2005;20:1320-8. https://doi.org/10.1093/ndt/gfh837.

[85] Birn H, Christensen EI. Renal albumin absorption in physiology and pathology. Kidney Int 2006;69:440-9. https://doi.org/10.1038/sj.ki.5000141.

[86] Ruiz-Ortega M, Esteban V, Rupérez M, Sánchez-López E, Rodríguez-Vita J, Carvajal G, et al. Renal and vascular hypertension-induced inflammation: role of angiotensin II. Curr Opin Nephrol Hypertens 2006;15:159-66. https://doi.org/10.1097/01.mnh.0000203190.34643.d4.

[87] Huang Y, Wongamorntham S, Kasting J, McQuillan D, Owens RT, Yu L, et al. Renin increases mesangial cell transforming growth factor-β1 and matrix proteins through receptor-mediated, angiotensin II-independent mechanisms. Kidney Int 2006;69:105-13. https://doi.org/10.1038/sj.ki.5000011.

[88] Rodríguez-Vita J, Sánchez-López E, Esteban V, Rupérez M, Egido J, Ruiz-Ortega M. Angiotensin II Activates the Smad Pathway in Vascular Smooth Muscle Cells by a Transforming Growth Factor-β–Independent Mechanism. Circulation 2005;111:2509-17. https://doi.org/10.1161/01.CIR.0000165133.84978.E2.

[89] Abrahamsen CT, Pullen MA, Schnackenberg CG, Grygielko ET, Edwards RM, Laping NJ, et al. Effects of Angiotensins II and IV on Blood Pressure, Renal Function, and PAI-1 Expression in the Heart and Kidney of the Rat. Pharmacology 2002;66:26-30. https://doi.org/10.1159/000063252.

[90] Kalluri R, Neilson EG. Epithelial-mesenchymal transition and its implications for fibrosis. J Clin Invest 2003;112:1776-84. https://doi.org/10.1172/JCI200320530.

[91] Zhan Y. Ets-1 is a critical regulator of Ang II-mediated vascular inflammation and remodeling. J Clin Invest 2005;115:2508-16. https://doi.org/10.1172/JCI24403.

[92] Wolf G, Bohlender J, Bondeva T, Roger T, Thaiss F, Wenzel UO. Angiotensin II Upregulates Toll-Like Receptor 4 on Mesangial Cells. J Am Soc Nephrol 2006;17:1585-93. https://doi.org/10.1681/ASN.2005070699.

[93] Jankowski V, Vanholder R, van der Giet M, Henning L, Tölle M, Schönfelder G, et al. Detection of Angiotensin II in Supernatants of Stimulated Mononuclear Leukocytes by Matrix-Assisted Laser Desorption Ionization Time-of-Flight/Time-of-Flight Mass Analysis. Hypertension 2005;46:591-7. https://doi.org/10.1161/01.HYP.0000177436.09733.d4.

[94] Anderson S, Rennke HG, Brenner BM. Therapeutic advantage of converting enzyme inhibitors in arresting progressive renal disease associated with systemic hypertension in the rat. J Clin Invest 1986;77:1993-2000. https://doi.org/10.1172/JCI112528.

[95] Rüster C, Wolf G. Renin-Angiotensin-Aldosterone System and Progression of Renal Disease. J Am Soc Nephrol 2006;17:2985-91. https://doi.org/10.1681/ASN.2006040356.

[96] Hollenberg NK, Price DA, Fisher NDL, Lansang MC, Perkins B, Gordon MS, et al. Glomerular hemodynamics and the renin-angiotensin system in patients with type 1 diabetes mellitus. Kidney Int 2003;63:172-8. https://doi.org/10.1046/j.1523-1755.2003.00701.x.

[97] Kobori H, Alper AB, Shenava R, Katsurada A, Saito T, Ohashi N, et al. Urinary Angiotensinogen as a Novel Biomarker of the Intrarenal Renin-Angiotensin System Status in Hypertensive Patients. Hypertension 2009;53:344-50. https://doi.org/10.1161/HYPERTENSIONAHA.108.123802.

[98] Yamamoto T, Nakagawa T, Suzuki H, Ohashi N, Fukasawa H, Fujigaki Y, et al. Urinary Angiotensinogen as a Marker of Intrarenal Angiotensin II Activity Associated with Deterioration of Renal Function in Patients with Chronic Kidney Disease. J Am Soc Nephrol 2007;18:1558-65. https://doi.org/10.1681/ASN.2006060554.

[99] Wolf G, Wenzel UO. Angiotensin II and Cell Cycle Regulation. Hypertension 2004;43:693-8. https://doi.org/10.1161/01.HYP.0000120963.09029.ca.

[100] Lakshmanan AP, Thandavarayan RA, Palaniyandi SS, Sari FR, Meilei H, Giridharan V V., et al. Modulation of AT-1R/CHOP-JNK-Caspase12 pathway by olmesartan treatment attenuates ER stress-induced renal apoptosis in streptozotocin-induced diabetic mice. Eur J Pharm Sci 2011;44:627-34. https://doi.org/10.1016/j.ejps.2011.10.009.

[101] Ha T-S, Park H-Y, Seong S-B, Ahn HY. Angiotensin II induces endoplasmic reticulum stress in podocyte, which would be further augmented by PI3-kinase inhibition. Clin Hypertens 2015;21:13. https://doi.org/10.1186/s40885-015-0018-5.

[102] Cardoso VG, Gonçalves GL, Costa-Pessoa JM, Thieme K, Lins BB, Casare FAM, et al. Angiotensin II-induced podocyte apoptosis is mediated by endoplasmic reticulum stress/PKC-δ/p38 MAPK pathway activation and trough increased Na+/H+ exchanger isoform 1 activity. BMC Nephrol 2018;19:179. https://doi.org/10.1186/s12882-018-0968-4.

[103] Raiden S, Nahmod K, Nahmod V, Semeniuk G, Pereira Y, Alvarez C, et al. Nonpeptide Antagonists of AT1 Receptor for Angiotensin II Delay the Onset of Acute Respiratory Distress Syndrome. J Pharmacol Exp Ther 2002;303:45-51. https://doi.org/10.1124/jpet.102.037382.

[104] Chen M, Chen C, Yuan X, Chen X, Zheng F, Shao L, et al. Angiotensin II aggravates lipopolysaccharide induced human pulmonary microvascular endothelial cells permeability in high glucose status. Endocr J 2018;65:717-25. https://doi.org/10.1507/endocrj.EJ17-0477.

[105] Zhang H, Penninger JM, Li Y, Zhong N, Slutsky AS. Angiotensin-converting enzyme 2 (ACE2) as a SARS-CoV-2 receptor: molecular mechanisms and potential therapeutic target. Intensive Care Med 2020;46:586-90. https://doi.org/10.1007/s00134-020-05985-9.

[106] Gurwitz D. Angiotensin receptor blockers as tentative SARS-CoV-2 therapeutics. Drug Dev Res 2020;81:537-40. https://doi.org/10.1002/ddr.21656.

[107] Tao W, Li P-S, Xu G, Luo Y, Shu Y-S, Tao Y-Z, et al. Soluble Epoxide Hydrolase Plays a Vital Role in Angiotensin II-Induced Lung Injury in Mice. SHOCK 2018;50:589-94. https://doi.org/10.1097/SHK.0000000000001067.

[108] Chen C, Zhang Z, Li Z, Zhang F, Peng M, Chen Y, et al. Losartan attenuates microvascular permeability in mechanical ventilator-induced lung injury in diabetic mice. Mol Biol Rep 2014;41:809-14. https://doi.org/10.1007/s11033-013-2920-9.

[109] Henry C, Zaizafoun M, Stock E, Ghamande S, Arroliga AC, White HD.

Impact of angiotensin-converting enzyme inhibitors and statins on viral pneumonia. Baylor Univ Med Cent Proc 2018;31:419-23. https://doi.org/10.1080/08998280.2018.1499293.

[110] Gu H, Xie Z, Li T, Zhang S, Lai C, Zhu P, et al. Angiotensin-converting enzyme 2 inhibits lung injury induced by respiratory syncytial virus. Sci Rep 2016;6:19840. https://doi.org/10.1038/srep19840.

[111] Zou Z, Yan Y, Shu Y, Gao R, Sun Y, Li X, et al. Angiotensin-converting enzyme 2 protects from lethal avian influenza A H5N1 infections. Nat Commun 2014;5:3594. https://doi.org/10.1038/ncomms4594.

[112] Yang P, Gu H, Zhao Z, Wang W, Cao B, Lai C, et al. Angiotensin-converting enzyme 2 (ACE2) mediates influenza H7N9 virus-induced acute lung injury. Sci Rep 2015;4:7027. https://doi.org/10.1038/srep07027.

[113] Yan Y, Liu Q, Li N, Du J, Li X, Li C, et al. Angiotensin II receptor blocker as a novel therapy in acute lung injury induced by avian influenza A H5N1 virus infection in mouse. Sci China Life Sci 2015;58:208-11. https://doi.org/10.1007/s11427-015-4814-7.

[114] Li W, Moore MJ, Vasilieva N, Sui J, Wong SK, Berne MA, et al. Angiotensin-converting enzyme 2 is a functional receptor for the SARS coronavirus. Nature 2003;426:450-4. https://doi.org/10.1038/nature02145.

[115] Heurich A, Hofmann-Winkler H, Gierer S, Liepold T, Jahn O, Pohlmann S. TMPRSS2 and ADAM17 Cleave ACE2 Differentially and Only Proteolysis by TMPRSS2 Augments Entry Driven by the Severe Acute Respiratory Syndrome Coronavirus Spike Protein. J Virol 2014;88:1293-307. https://doi.org/10.1128/JVI.02202-13.

[116] Kuba K, Imai Y, Rao S, Gao H, Guo F, Guan B, et al. A crucial role of angiotensin converting enzyme 2 (ACE2) in SARS coronavirus–induced lung injury. Nat Med 2005;11:875-9. https://doi.org/10.1038/nm1267.

[117] Itoyama S, Keicho N, Quy T, Phi NC, Long HT, Ha LD, et al. ACE1 polymorphism and progression of SARS. Biochem Biophys Res Commun 2004;323:1124-9. https://doi.org/10.1016/j.bbrc.2004.08.208.

[118] Kong SL, Chui P, Lim B, Salto-Tellez M. Elucidating the molecular physiopathology of acute respiratory distress syndrome in severe acute respiratory syndrome patients. Virus Res 2009;145:260-9. https://doi.org/10.1016/j.virusres.2009.07.014.

[119] Hoffmann M, Kleine-Weber H, Schroeder S, Krüger N, Herrler T, Erichsen S, et al. SARS-CoV-2 Cell Entry Depends on ACE2 and TMPRSS2 and Is Blocked by a Clinically Proven Protease Inhibitor. Cell 2020;181:271-280.e8. https://doi.org/10.1016/j.cell.2020.02.052.

[120] Zhou P, Yang X-L, Wang X-G, Hu B, Zhang L, Zhang W, et al. A pneumonia outbreak associated with a new coronavirus of probable bat origin. Nature 2020;579:270-3. https://doi.org/10.1038/s41586-020-2012-7.

[121] Xu X, Chen P, Wang J, Feng J, Zhou H, Li X, et al. Evolution of the novel coronavirus from the ongoing Wuhan outbreak and modeling of its spike protein for risk of human transmission. Sci China Life Sci 2020;63:457-60. https://doi.org/10.1007/s11427-020-1637-5.

[122] Letko M, Marzi A, Munster V. Functional assessment of cell entry and receptor usage for SARS-CoV-2 and other lineage B betacoronaviruses. Nat Microbiol 2020;5:562-9. https://doi.org/10.1038/s41564-020-0688-y.

[123] Li J, Liu W. Puzzle of highly pathogenic human coronaviruses

(2019-nCoV). Protein Cell 2020;11:235-8. https://doi.org/10.1007/s13238-020-00693-y.

[124] Wan Y, Shang J, Graham R, Baric RS, Li F. Receptor Recognition by the Novel Coronavirus from Wuhan: an Analysis Based on Decade-Long Structural Studies of SARS Coronavirus. J Virol 2020;94. https://doi.org/10.1128/JVI.00127-20.

[125] Liu Y, Yang Y, Zhang C, Huang F, Wang F, Yuan J, et al. Clinical and biochemical indexes from 2019-nCoV infected patients linked to viral loads and lung injury. Sci China Life Sci 2020;63:364-74. https://doi.org/10.1007/s11427-020-1643-8.

[126] Chen N, Zhou M, Dong X, Qu J, Gong F, Han Y, et al. Epidemiological and clinical characteristics of 99 cases of 2019 novel coronavirus pneumonia in Wuhan, China: a descriptive study. Lancet 2020;395:507-13. https://doi.org/10.1016/S0140-6736(20)30211-7.

[127] Pati A, Mahto H, Padhi S, Panda AK. ACE deletion allele is associated with susceptibility to SARS-CoV-2 infection and mortality rate: An epidemiological study in the Asian population. Clin Chim Acta 2020;510:455-8. https://doi.org/10.1016/j.cca.2020.08.008.

[128] Labandeira-Garcia JL, Rodríguez-Perez AI, Garrido-Gil P, Rodriguez-Pallares J, Lanciego JL, Guerra MJ. Brain Renin-Angiotensin System and Microglial Polarization: Implications for Aging and Neurodegeneration. Front Aging Neurosci 2017;9. https://doi.org/10.3389/fnagi.2017.00129.

[129] Takane K, Hasegawa Y, Lin B, Koibuchi N, Cao C, Yokoo T, et al. Detrimental Effects of Centrally Administered Angiotensin II are Enhanced in a Mouse Model of Alzheimer Disease Independently of Blood Pressure. J Am Heart Assoc 2017;6. https://doi.org/10.1161/JAHA.116.004897.

[130] Wright JW, Harding JW. The brain renin–angiotensin system: a diversity of functions and implications for CNS diseases. Pflügers Arch - Eur J Physiol 2013;465:133-51. https://doi.org/10.1007/s00424-012-1102-2.

[131] Parihar M., Hemnani T. Alzheimer's disease pathogenesis and therapeutic interventions. J Clin Neurosci 2004;11:456-67. https://doi.org/10.1016/j.jocn.2003.12.007.

[132] Murakami T, Paitel E, Kawarabayashi T, Ikeda M, Chishti MA, Janus C, et al. Cortical Neuronal and Glial Pathology in TgTauP301L Transgenic Mice. Am J Pathol 2006;169:1365-75. https://doi.org/10.2353/ajpath.2006.051250.

[133] Akiyama H, Barger S, Barnum S, Bradt B, Bauer J, Cole GM, et al. Inflammation and Alzheimer's disease. Neurobiol Aging 2000;21:383-421. https://doi.org/10.1016/S0197-4580(00)00124-X.

[134] Kumar A, Dhull DK, Mishra PS. Therapeutic potential of mGluR5 targeting in Alzheimer's disease. Front Neurosci 2015;9. https://doi.org/10.3389/fnins.2015.00215.

[135] Praticò D, Sung S. Lipid Peroxidation and Oxidative imbalance: Early functional events in Alzheimer's disease. J Alzheimer's Dis 2004;6:171-5. https://doi.org/10.3233/JAD-2004-6209.

[136] Manczak M, Anekonda TS, Henson E, Park BS, Quinn J, Reddy PH. Mitochondria are a direct site of Aβ accumulation in Alzheimer's disease neurons: implications for free radical generation and oxidative damage in disease progression. Hum Mol Genet 2006;15:1437-49. https://doi.org/10.1093/hmg/ddl066.

[137] Mascolo A, Sessa M, Scavone C, De Angelis A, Vitale C, Berrino L, et al. New and old roles of the peripheral and brain renin–angiotensin–aldosterone system (RAAS): Focus on cardiovascular and neurological diseases. Int J Cardiol 2017;227:734-42. https://doi.org/10.1016/j.ijcard.2016.10.069.

[138] Rygiel K. Can angiotensin-converting enzyme inhibitors impact cognitive decline in early stages of Alzheimer's disease? An overview of research evidence in the elderly patient population. J Postgrad Med 2016;62:242. https://doi.org/10.4103/0022-3859.188553.

[139] Ohrui T, Tomita N, Sato-Nakagawa T, Matsui T, Maruyama M, Niwa K, et al. Effects of brain-penetrating ACE inhibitors on Alzheimer disease progression. Neurology 2004;63:1324-5. https://doi.org/10.1212/01.WNL.0000140705.23869.E9.

[140] Bodiga VL, Bodiga S. Renin Angiotensin System in Cognitive Function and Dementia. Asian J Neurosci 2013;2013:1-18. https://doi.org/10.1155/2013/102602.

[141] Kehoe PG, Wong S, AL Mulhim N, Palmer LE, Miners JS. Angiotensin-converting enzyme 2 is reduced in Alzheimer's disease in association with increasing amyloid-β and tau pathology. Alzheimers Res Ther 2016;8:50. https://doi.org/10.1186/s13195-016-0217-7.

[142] Uekawa K, Hasegawa Y, Senju S, Nakagata N, Ma M, Nakagawa T, et al. Intracerebroventricular Infusion of Angiotensin-(1-7) Ameliorates Cognitive Impairment and Memory Dysfunction in a Mouse Model of Alzheimer's Disease. J Alzheimer's Dis 2016;53:127-33. https://doi.org/10.3233/JAD-150642.

[143] Prusty S, Sahu P, Subudhi B. Angiotensin Mediated Oxidative Stress and Neuroprotective Potential of Antioxidants and AT1 Receptor Blockers. Mini-Reviews Med Chem 2017;17:518-28. https://doi.org/10.2174/1389557516666161025094539.

[144] Gebre AK, Altaye BM, Atey TM, Tuem KB, Berhe DF. Targeting Renin–Angiotensin System Against Alzheimer's Disease. Front Pharmacol 2018;9. https://doi.org/10.3389/fphar.2018.00440.

[145] Ali MRA-A, Abo-Youssef AMH, Messiha BAS, Khattab MM. Tempol and perindopril protect against lipopolysaccharide-induced cognition impairment and amyloidogenesis by modulating brain-derived neurotropic factor, neuroinflammation and oxido-nitrosative stress. Naunyn Schmiedebergs Arch Pharmacol 2016;389:637-56. https://doi.org/10.1007/s00210-016-1234-6.

[146] Goel R, Bhat SA, Hanif K, Nath C, Shukla R. Perindopril Attenuates Lipopolysaccharide-Induced Amyloidogenesis and Memory Impairment by Suppression of Oxidative Stress and RAGE Activation. ACS Chem Neurosci 2016;7:206-17. https://doi.org/10.1021/acschemneuro.5b00274.

[147] Bild W, Hritcu L, Stefanescu C, Ciobica A. Inhibition of central angiotensin II enhances memory function and reduces oxidative stress status in rat hippocampus. Prog Neuro-Psychopharmacology Biol Psychiatry 2013;43:79-88. https://doi.org/10.1016/j.pnpbp.2012.12.009.

[148] Seifi B, Kadkhodaee M, Bakhshi E, Ranjbaran M, Zahmatkesh M, Sedaghat Z, et al. Angiotensin II in paraventricular nucleus contributes to sympathoexcitation in renal ischemia–reperfusion injury by AT1 receptor and oxidative stress. J Surg Res 2015;193:361-7. https://doi.org/10.1016/j.jss.2014.06.042.

[149] Erdi F, Keskin F, Esen H, Kaya B, Feyzioglu B, Kilinc I, et al. Telmisartan ameliorates oxidative stress and subarachnoid haemorrhage-induced cerebral vasospasm. Neurol Res 2016;38:224-31. https://doi.org/10.1080/01616412.2015.1105626.

[150] Barone E, Head E, Butterfield DA, Perluigi M. HNE-modified proteins in Down syndrome: Involvement in development of Alzheimer disease neuropathology. Free Radic Biol Med 2017;111:262-9. https://doi.org/10.1016/j.freeradbiomed.2016.10.508.

[151] Tota S, Kamat PK, Awasthi H, Singh N, Raghubir R, Nath C, et al. Candesartan improves memory decline in mice: Involvement of AT1 receptors in memory deficit induced by intracerebral streptozotocin. Behav Brain Res 2009;199:235-40. https://doi.org/10.1016/j.bbr.2008.11.044.

[152] Gao Q, Jiang T, Zhao H-R, Wu L, Tian Y-Y, Ou Z, et al. Activation of Autophagy Contributes to the Angiotensin II-Triggered Apoptosis in a Dopaminergic Neuronal Cell Line. Mol Neurobiol 2016;53:2911-9. https://doi.org/10.1007/s12035-015-9177-3.

[153] Zawada WM, Mrak RE, Biedermann J, Palmer QD, Gentleman SM, Aboud O, et al. Loss of angiotensin II receptor expression in dopamine neurons in Parkinson's disease correlates with pathological progression and is accompanied by increases in Nox4- and 8-OH guanosine-related nucleic acid oxidation and caspase-3 activation. Acta Neuropathol Commun 2015;3:9. https://doi.org/10.1186/s40478-015-0189-z.

[154] Zubenko GS, Volicer L, Direnfeld LK, Freeman M, Langlais PJ, Nixon RA. Cerebrospinal fluid levels of angiotensin-converting enzyme in Alzheimer's disease, Parkinson's disease and progressive supranuclear palsy. Brain Res 1985;328:215-21. https://doi.org/10.1016/0006-8993(85)91032-7.

[155] Kingwell E, Marriott JJ, Jetté N, Pringsheim T, Makhani N, Morrow SA, et al. Incidence and prevalence of multiple sclerosis in Europe: a systematic review. BMC Neurol 2013;13:128. https://doi.org/10.1186/1471-2377-13-128.

[156] Kamm CP, Uitdehaag BM, Polman CH. Multiple Sclerosis: Current Knowledge and Future Outlook. Eur Neurol 2014;72:132-41. https://doi.org/10.1159/000360528.

[157] Platten M, Youssef S, Hur EM, Ho PP, Han MH, Lanz T V., et al. Blocking angiotensin-converting enzyme induces potent regulatory T cells and modulates TH1- and TH17-mediated autoimmunity. Proc Natl Acad Sci 2009;106:14948-53. https://doi.org/10.1073/pnas.0903958106.

[158] Babaloo Z, Aliparasti MR, Babaiea F, Almasi S, Baradaran B, Farhoudi M. The role of Th17 cells in patients with relapsing-remitting multiple sclerosis: Interleukin-17A and interleukin-17F serum levels. Immunol Lett 2015. https://doi.org/10.1016/j.imlet.2015.01.001.

[159] Constantinescu CS, Goodman DBP, Grossman RI, Mannon LJ, Cohen JA. Serum Angiotensin-Converting Enzyme in Multiple Sclerosis. Arch Neurol 1997;54:1012-5. https://doi.org/10.1001/archneur.1997.00550200068012.

[160] Kawajiri M, Mogi M, Higaki N, Matsuoka T, Ohyagi Y, Tsukuda K, et al. Angiotensin-converting enzyme (ACE) and ACE2 levels in the cerebrospinal fluid of patients with multiple sclerosis. Mult Scler J 2009;15:262-5. https://doi.org/10.1177/1352458508097923.

Chapter 4

Renin Angiotensin Aldosterone System Functions in Renovascular Hypertension

Jose A. Gomez

Abstract

The renin angiotensin aldosterone system (RAAS) plays a key function in renovascular hypertension induced by renal artery stenosis (RAS). RAS causes a decrease in renal perfusion in the stenosed kidney which in turn stimulates renin the rate limiting enzyme in RAAS. This stimulation triggers a series of events starting with renin release leading to Ang II production, decrease in sodium excretion, increase sympathetic tone; all contributing to the development of renovascular hypertension. In RAS increase of superoxide reduce nitric oxide in the afferent arteriole increasing vasoconstriction and a marked decrease in glomerular filtration rate. In renovascular hypertension prostaglandins mediate renin release in the stenosed kidney. Targeting different RAAS components is part of the therapy for renovascular hypertension, with other options including renal nerves denervation and revascularization. Different clinical studies had explored revascularization, RAAS blocking and renal nerves denervation as a therapy. We will discuss organ, cellular and molecular components of this disease.

Keywords: Renin angiotensin aldosterone system, renovascular hypertension, renin, renal nerves, oxidative stress

1. Introduction

Renal artery stenosis (RAS) is a common condition in patients suffering from atherosclerosis and fibromuscular dysplasia [1–6], with an overall prevalence disease rate of 15.4% [4]. Progression to severe stenosis is well documented and leads to hypertension and kidney damage [7–9]. Clinically, renovascular hypertension is one the most important causes of secondary hypertension and kidney damage. In patients with RAS, 65% are hypertensive and 26.5% suffer kidney failure [4, 6]. Advancement to end stage renal disease is known to increase cardiovascular events [10]. The clinical trials Angioplasty and Stenting for Renal Artery Lesions (ASTRAL) [11], and Cardiovascular Outcomes in Renal Atherosclerotic Lesions (CORAL) [12] targeted renal vascularization to improve disease outcomes but failed to show any improvement in renal function, cardiovascular events or mortality [11, 12]. Furthermore, prospective studies in ASTRAL and CORAL concluded that 15-22% of patients suffering from renovascular disease will progress to renal "end point" within 3 to 4 years [13]. The NHLBI Cardiovascular Health Study used a non-invasive screen and found that 6.8% elderly patients (both African American

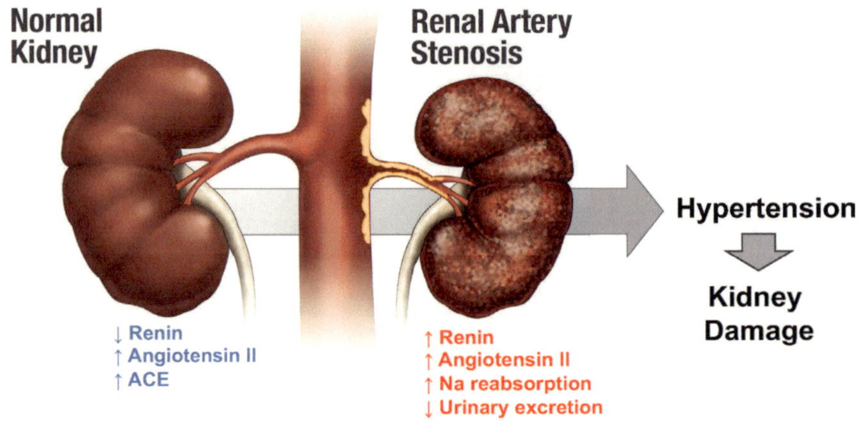

Figure 1.
Renin Angiotensin Aldosterone System (RAAS) key role in renal artery stenosis (RAS) induction of renovascular hypertension and kidney damage. Deterioration of renal perfusion in the stenosed kidney cause a decrease in renal pressure which in turn stimulates RAAS. This stimulation triggers a series of events starting with renin release leading to angiotensin II production; decrease in sodium excretion, increase sympathetic tone; ending in hypertension.

and white) had more than 60% RASten or renal artery occlusion [14, 15]. The renin angiotensin aldosterone system (RAAS) plays a key role in hypertension, with renin recognized as the driver of renovascular hypertension (**Figure 1**). In humans, plasma renin activity (PRA) is used as biomarker for the activation of RAAS in hypertension and in patients with atherosclerotic RAS, high PRA is associated with increased risk for cardiovascular events and high mortality [16]. These suggest an important function for RAAS in renovascular hypertension onset and the need to target different components of RAAS for therapy.

2. Renin angiotensin aldosterone system function in renal artery stenosis

Renal artery stenosis causes a decrease in renal perfusion in the stenosed kidney which in turn stimulates RAAS. This stimulation triggers a series of events starting with renin release leading to angiotensin II (Ang II) production, decrease in sodium excretion, increase sympathetic tone; all contributing to the development of hypertension (**Figure 1**) [17, 18]. When there is a need for renin expression and release, the number of renin expressing cells increase a process known as Juxtaglomerular (JG) cell recruitment [19–24] involving the trans differentiation of vascular smooth muscle cells into renin expressing cells along the afferent arteriole [20, 21, 23]. JG cell recruitment is well documented in this model [25–27]. Activation of the renal baroreceptor in RAS causes renovascular hypertension through RAAS activation [28]. In uni- and bi-lateral RAS aldosterone levels are upregulated [29–32]. Moreover, in renovascular hypertension prostaglandins mediate renin release in the stenosed kidney [33–36], and catecholamines mediated by an increase in cAMP and activation of protein kinase A (PKA) [37–39]. Decrease renal perfusion cause a decline in renal function and increase kidney injury [40, 41]. This decrease in renal function starts with endothelial damage, decrease in nitric oxide and increase in vasoconstrictors and oxidative species [42]. Reactive oxidative stress (ROS) increase renal vascular tone, tubuloglomerular feedback, and endothelial disfunction decreasing glomerular filtration rate [43].

Successful treatments for hypertension such as angiotensin converting enzyme (ACE) inhibitors and angiotensin receptor blockers (ARBs) alleviate hypertension, but need close examining for kidney failure and hyperkalemia [4]. Aliskiren, a direct renin inhibitor, may still be a potential option for the treatment of high blood pressure in some forms of hypertension such as chronic kidney disease (CKD) and renovascular hypertension [44]. In a clinical study, aliskiren combined with olmesartan reduced proteinuria by about 40% from baseline in patients with CKD with persistent proteinuria [45]. In non-diabetic CKD patients, aliskiren combined with ARBs, safely reduced proteinuria and attenuated the decline in glomerular filtration rate (GFR) [46]. These results indicate that a complete treatment of renal artery stenosis induced renovascular hypertension and kidney damage may need targeting both the angiotensin II-dependent and the Ang II-independent arms of RAAS.

Renal artery stenosis is common in diabetic patients placing them at higher risk of end organ damage causing end stage renal disease [9, 47–49]. In older patients, RAS is the most common problem of end stage renal failure [50]. In RAS renin is recognized as the disease driver [6, 16, 51–54]. RAS is common in atherosclerotic patients and caused hypertension, oxidative stress, and kidney damage [7, 9]. Increased oxidative stress has been reported in humans as well as in two kidney one clip (2K1C) animal model and other hypertensive animal models [24, 55–60]. Changes in renal perfusion activate RAAS and increase the sympathetic activity of the afferent renal nerves contributing to renovascular hypertension and end-stage renal disease during RAS [61]. In the 2K1C model renal denervation decreases hypertension [62, 63]. Clinical trials (Renal Denervation in Patients With Refractory Hypertension (HTN-1) (Symplicity HTN-1), Renal Denervation in Patients With Uncontrolled Hypertension (Symplicity HTN-2), The Renal Denervation for Hypertension (DENERHTN), and Catheter-based renal denervation in patients with uncontrolled hypertension in the absence of antihypertensive medications (SPYRAL)) report that using renal denervation as therapy for hypertension has good outcomes [64–67]. The therapeutic effects of renal denervation have been attributed to removal of sympathetic efferent and/or afferent fibers [68]. Renin secretion is stimulated by renal efferent nerves, which also stimulate tubular sodium reabsorption [62] without perturbations to glomerular filtration rate or albumin urinary secretion [69]. These indicates that initially, renal artery stenosis induces RAAS and in later stages other organs involved in blood pressure homeostasis are involved in the induction of renovascular hypertension such as renal nerves and adrenal gland.

3. Central nervous system input in renal artery stenosis

Different experimental models of hypertension showed the crucial role play by the central nervous system (CNS) in this disease. Specifically, sympathetic efferent outflow augments during hypertension. It has been shown that both Ang II and aldosterone actions are mediated by the CNS [70, 71]. In experimental models of hypertension, ablation of the forebrain surrounding the anteroventral third cerebral ventricle (AV3V) inhibited hypertension [72, 73]. In the CNS the AV3V contains the median preoptic eminence, the organum vasculosum of the lateral terminalis, and the preoptic periventricular nucleus [74]. This forebrain region is responsible for cardiovascular regulation, and includes the subfornical organ, the organum vasculosum of the lamina terminalis, which are circumventricular organs lacking a blood-brain barrier [75]. Production of ROS in these brain regions strongly influences blood pressure [76]. Several reports showed that actions on these brain regions are responsible for Ang II hypertension and increase oxidative stress with NADPH oxidase playing a key role [77–80]. Renal vasculature and tubular segments are controlled by the efferent sympathetic renal nerves and promote arteriolar vasoconstriction and renin release and increases sodium reabsorption [81]. In the afferent arterioles Ang II activates the alpha1 adrenergic receptor, which increases oxidative stress and constriction of the afferent arterioles, reducing renal blood flow [82]. Contrary, activation of the b1-adrenergic receptor activation inhibits ROS generation promoting vasodilation [83]. In different hypertension animal models renal denervation inhibit the induction of hypertension, showing that ablation of renal efferent induction of ROS is important in hypertension development [84, 85]. These data indicate that oxidative stress control efferent and afferent renal nerve actions in the development of hypertension.

Renal artery stenosis activates RAAS and increases the activity of the afferent renal nerves resulting in hypertension and end-stage renal disease [61]. It is known that in the 2K1C model renal denervation decreases hypertension [62, 63]. Removal of sympathetic efferent and/or afferent fibers controls hypertension [68], and the renal efferent nerves stimulate renin secretion and tubular sodium reabsorption [62]. During renal artery stenosis, there is an increase in Neutrophil Oxidase Factor p47 (p47phox) and p67phox [86–88]. Furthermore, in renal artery stenosis generation of ROS induced renal damage [88, 89], with the main source of ROS being NADPH oxidase [90, 91].

In the induction of renovascular hypertension, the renal nerves as well as the renin angiotensin aldosterone system activation cause the increase in blood pressure and dysregulation of sodium secretion, with renal denervation alleviating the central nerve system input decreasing blood pressure.

4. Oxidative stress in renal artery stenosis

Oxidative stress in the kidney and vasculature contribute to hypertension development. NADPH oxidase is a major source of oxidative stress in mammalian cells [75]. Most of the renal cells express NADPH oxidase and there are several stimuli that cause its activation leading to organ injury and hypertension development [75, 92, 93]. Reactive oxygen species (ROS) produced by NADPH oxidase in the kidney cause vasoconstriction and organ injury. Specifically, increase of superoxide reduces nitric oxide (NO) in the afferent arteriole increasing vasoconstriction and a marked decrease in GFR. In rabbits, Ang II-induced hypertension increase the p22phox subunit of NADPH oxidase causing endothelial dysfunction in the afferent arteriole [94]. Moreover, in spontaneous hypertensive rats, superoxide is generated in the afferent arteriole in response to endothelin-1 (ET-1) [95, 96]. Podocytes are important components of the renal filtration system. Dahl salt-sensitive rats had increase glomerular expression of p22phos and NOX2 that increases oxidative stress causing podocyte injury, glomerular sclerosis and proteinuria, with the antioxidant tempol (4-Hydroxy-TEMPO) correcting this glomerular injury [97, 98]. Plasminogen causes podocyte injury through stimulation of NOX2 and NOX4 expression [99], Ang II stimulates ROS generation in the mitochondria stimulating autophagy [100], Ang II-induced ROS production caused glomerulosclerosis [101], and oxidative stress disrupts nephrin – caveolin-1 crosstalk in podocytes disrupting of glomerular filtration barrier [102]. In the vasculature, increased oxidative stress causes hypertension in different animal models [103–108]. During renal artery stenosis, generation of ROS is recognized as the main mechanism of renal damage [88, 89, 109, 110] with the activation of NADPH oxidase as the source of ROS [90, 91], and associated with an increase in p47phox and p67phox [19, 86–88].

It is important to recognize that renal artery stenosis increase the production of reactive oxygen species leading to renal damage. ROS production influences not only organ damage but also contributes to the increase in blood pressure.

In the therapy of this disease multiple molecules are involved leading to increases in oxidative stress, blood pressure and renal injury and all start with the activation of the renin angiotensin aldosterone system.

5. Angiotensin II dependent and independent action in renal artery stenosis

In renal artery stenosis induction of renovascular hypertension, renin is recognized a key molecule, and as such in the therapy of renovascular hypertension Angiotensin Converting Enzyme (ACE) inhibitors and Angiotensin Receptor blockers (ARBs) are used [4]. Moreover, sympathetic nervous systems action in the kidney promotes renin secretion through renal efferent nerves, which also stimulate tubular sodium reabsorption [62], and in the 2K1C model denervation inhibit the onset of hypertension [62, 63]. Renal artery stenosis causes renovascular hypertension, which is associated with deterioration of kidney function [20]. Reduction in renal flow is recognize as a source of hypoxia during renovascular hypertension [21]. Arterial stenosis causes thrombosis, and ischemia in renovascular hypertension [22]. During renal artery stenosis generation of ROS is recognized as the main mechanism of renal damage [88, 89], causing increased in vasoconstrictors, cell death and decrease in the activity of nitric oxide [109, 110]. A swine model of renal artery stenosis presented an increase in ROS, renal and cardiac damage [23, 86–89, 111–113]. In renal artery stenosis activation of RAAS increase ROS generating by the activation of NADPH oxidase [90, 91], associated with is an increase in p47phox and p67phox [86–88]. Phosphorylation of p47phox by PKC is a key step in NADPH oxidase activation [114–118]. Hypertension is associated with PKC activation and increase oxidative stress [119], which caused endothelial nitric oxide synthase (eNOS) disfunction and uncoupling producing ROS instead of NO. This uncoupling is a key mechanism for endothelial dysfunction in angiotensin II-induced hypertension [120–122]. Increase in NOX2 activity requires increase NOX2 expression and p47phox association and activation of NOX2 [19]. Furthermore, increase in oxidative stress is well documented in 2K1C model [55–59, 123, 124]. All the actions mentioned above are Ang II mediated.

New evidence places (pro)renin receptor (PRR) as an effector molecule in the Ang II-independent RAAS [125]. PRR binds both renin and prorenin [125–129]. There is an association of PRR with different pathophysiology of diseases [130–135]. PRR binds renin causing an increase in Ang I [125] and it can activate prorenin by promoting a conformational change [125–129]. PRR mRNA is expressed in different organs such as kidney, heart, brain, eye, adipose tissue and vascular SMCs [125, 134], It has been proposed that PRR activates the Ang II-independent RAAS with tissue specificity [136]. My laboratory and others are uncovering new functions of the Ang II independent pathway in blood pressure, oxidative stress and organ damage. New studies will define the relevance of this arm of RAAS and possible define new molecular targets for therapy.

6. Concluding remarks and future perspectives

In the definition of the molecular pathways involved in the development of renovascular hypertension, the Goldblatt two kidney one clip animal model has been critical. This animal mode has been extensively used with different animals all showing that renal artery stenosis strongly stimulates renin overexpression and release promoting renovascular hypertensions and kidney injury. In renovascular hypertension renin is key and promotes the increase in Ang II leading to hypertension. Renin being the rate limiting step in the production of Ang II in RAAS, has been investigated as a possible target for the therapy. However, the main therapies used are angiotensin converting enzyme inhibitors and angiotensin receptor blockers. Direct renin inhibition by aliskiren, is potential therapy for hypertension in chronic kidney disease (CKD) and renovascular hypertension. Combination of aliskiren with olmesartan in the clinic, reduced proteinuria in patients with CKD with persistent proteinuria. In non-diabetic CKD patients, aliskiren combined with ARBs, reduced proteinuria and protected from the decline in glomerular filtration rate. We have shown here clinical and research data that indicates the during renal artery stenosis induced renovascular hypertension RAAS is activated and play a critical role in this pathology. It is important that a complete treatment of renovascular hypertension may need targeting both the angiotensin II-dependent and the Ang II-independent arms of RAAS.

Acknowledgements

We apologize to colleagues whose work has not been included in this chapter due to space limitations. This work was supported by NIH Grants: NHLBI Research Scientist Development Grant (1K01HL135461).

Author details

Jose A. Gomez
Department of Medicine, Clinical Pharmacology Division, Vanderbilt University Medical Center, Nashville, TN, United States

*Address all correspondence to: jose.a.gomez@vumc.org

References

[1] Jean WJ, al-Bitar I, Zwicke DL, Port SC, Schmidt DH, Bajwa TK. High incidence of renal artery stenosis in patients with coronary artery disease. Cathet Cardiovasc Diagn 1994;32:8-10.

[2] Olin JW, Melia M, Young JR, Graor RA, Risius B. Prevalence of atherosclerotic renal artery stenosis in patients with atherosclerosis elsewhere. Am J Med 1990;88:46N-51N.

[3] Silva JA. Evaluation and approach to treatment of renal artery stenosis in patients with diabetic nephropathy. Curr Diab Rep 2008;8:494-8.

[4] de Mast Q, Beutler JJ. The prevalence of atherosclerotic renal artery stenosis in risk groups: a systematic literature review. J Hypertens 2009;27:1333-40.

[5] Cooper EL, Xie Y, Nguyen H et al. Early Rapid Decline in Kidney Function in Medically Managed Patients With Atherosclerotic Renal Artery Stenosis. J Am Heart Assoc 2019;8:e012366.

[6] Textor SC, Lerman L. Renovascular hypertension and ischemic nephropathy. Am J Hypertens 2010;23:1159-69.

[7] Schreiber MJ, Pohl MA, Novick AC. The natural history of atherosclerotic and fibrous renal artery disease. Urol Clin North Am 1984;11:383-92.

[8] Crowley JJ, Santos RM, Peter RH et al. Progression of renal artery stenosis in patients undergoing cardiac catheterization. Am Heart J 1998;136:913-8.

[9] Caps MT, Perissinotto C, Zierler RE et al. Prospective study of atherosclerotic disease progression in the renal artery. Circulation 1998;98:2866-72.

[10] Al-Suraih M, Grande JP. Management of renal artery stenosis: What does the experimental evidence tell us? World J Cardiol 2014;6:855-60.

[11] Investigators A, Wheatley K, Ives N et al. Revascularization versus medical therapy for renal-artery stenosis. N Engl J Med 2009;361:1953-62.

[12] Cooper CJ, Murphy TP, Cutlip DE et al. Stenting and medical therapy for atherosclerotic renal-artery stenosis. N Engl J Med 2014;370:13-22.

[13] Textor SC. Renal Arterial Disease and Hypertension. Med Clin North Am 2017;101:65-79.

[14] Hansen KJ, Edwards MS, Craven TE et al. Prevalence of renovascular disease in the elderly: a population-based study. J Vasc Surg 2002;36:443-51.

[15] Weber BR, Dieter RS. Renal artery stenosis: epidemiology and treatment. Int J Nephrol Renovasc Dis 2014;7:169-81.

[16] Daidoji H, Tamada Y, Suzuki S et al. Plasma Renin Activity Predicts the Improvement in Resistant Hypertension after Percutaneous Transluminal Renal Artery Angioplasty. Intern Med 2016;55:3421-3426.

[17] An epidemiological approach to describing risk associated with blood pressure levels. Final report of the Working Group on Risk and High Blood Pressure. Hypertension 1985;7:641-51.

[18] Border WA, Noble NA. Interactions of transforming growth factor-beta and angiotensin II in renal fibrosis. Hypertension 1998;31:181-8.

[19] Brandes RP, Weissmann N, Schroder K. Nox family NADPH oxidases: Molecular mechanisms of activation. Free Radic Biol Med 2014;76:208-26.

[20] Plouin PF, Rossignol P, Bobrie G. Atherosclerotic renal artery stenosis:

to treat conservatively, to dilate, to stent, or to operate? J Am Soc Nephrol 2001;12:2190-6.

[21] Liu J, Wei Q, Guo C et al. Hypoxia, HIF, and Associated Signaling Networks in Chronic Kidney Disease. Int J Mol Sci 2017;18.

[22] Yu W, Liu-Bryan R, Stevens S, Damanahalli JK, Terkeltaub R. RAGE signaling mediates post-injury arterial neointima formation by suppression of liver kinase B1 and AMPK activity. Atherosclerosis 2012;222:417-25.

[23] Rajagopalan S, Duquaine D, King S, Pitt B, Patel P. Mineralocorticoid receptor antagonism in experimental atherosclerosis. Circulation 2002;105:2212-6.

[24] Saleem M., Saavedra-Sánchez L., Barturen-Larrea P., and Gomez J.A. The transcription factor Sox6 controls renin expression during renal artery stenosis. Kidney360 March 2021, 10.34067/KID.0002792020; DOI: https://doi.org/10.34067/KID.0002792020.

[25] Tomanek RJ, Schalk KA, Marcus ML, Harrison DG. Coronary angiogenesis during long-term hypertension and left ventricular hypertrophy in dogs. Circ Res 1989; 65:352-9.

[26] Braam B, Navar LG, Mitchell KD. Modulation of tubuloglomerular feedback by angiotensin II type 1 receptors during the development of Goldblatt hypertension. Hypertension 1995;25:1232-7.

[27] Gollan F, Richardson E, Goldblatt H. Hypertension in the systemic blood of animals with experimental renal hypertension. J Exp Med 1948;88: 389-400.

[28] Reinhold SW, Uihlein DC, Boger CA et al. Renin, endothelial NO synthase and endothelin gene expression in the 2kidney-1clip Goldblatt model of long-term renovascular hypertension. Eur J Med Res 2009;14:520-5.

[29] Kotliar C, Inserra F, Forcada P et al. Are plasma renin activity and aldosterone levels useful as a screening test to differentiate between unilateral and bilateral renal artery stenosis in hypertensive patients? J Hypertens 2010;28:594-601.

[30] Maxwell MH, Lupu AN, Viskoper RJ, Aravena LA, Waks UA. Mechanisms of hypertension during the acute and intermediate phases of the one-clip, two-kidney model in the dog. Circ Res 1977;40:I24-8.

[31] McAreavey D, Brown JJ, Cumming AM et al. Inverse relation of exchangeable sodium and blood pressure in hypertensive patients with renal artery stenosis. J Hypertens 1983;1:297-302.

[32] Takabatake T, Ohta H, Yamamoto Y et al. Effect of angiotensin blockade and converting enzyme inhibition on renovascular hypertension: comparison between unilateral and bilateral renal artery stenosis. Angiology 1987;38: 434-9.

[33] Imanishi M, Kawamura M, Akabane S et al. Aspirin lowers blood pressure in patients with renovascular hypertension. Hypertension 1989;14:461-8.

[34] Schricker K, Hamann M, Kaissling B, Kurtz A. Role of the macula densa in the control of renal renin gene expression in two-kidney/one-clip rats. Pflugers Arch 1994;427:42-6.

[35] Tuttle KR. Charting new territory by simulated modeling of a clinical trial. Clin J Am Soc Nephrol 2010;5:750-2.

[36] Wang JL, Cheng HF, Harris RC. Cyclooxygenase-2 inhibition decreases

renin content and lowers blood pressure in a model of renovascular hypertension. Hypertension 1999;34:96-101.

[37] Friis UG, Jensen BL, Sethi S, Andreasen D, Hansen PB, Skott O. Control of renin secretion from rat juxtaglomerular cells by cAMP-specific phosphodiesterases. Circ Res 2002;90: 996-1003.

[38] Schramm A, Schweda F, Sequeira-Lopez MLS, Hofmann F, Sandner P, Schlossmann J. Protein Kinase G Is Involved in Acute but Not in Long-Term Regulation of Renin Secretion. Front Pharmacol 2019;10:800.

[39] Sparks MA, Crowley SD, Gurley SB, Mirotsou M, Coffman TM. Classical Renin-Angiotensin system in kidney physiology. Compr Physiol 2014;4: 1201-28.

[40] Fava C, Minuz P, Patrignani P, Morganti A. Renal artery stenosis and accelerated atherosclerosis: which comes first? J Hypertens 2006;24:1687-96.

[41] Chade AR, Lerman A, Lerman LO. Kidney in early atherosclerosis. Hypertension 2005;45:1042-9.

[42] Brodsky SV, Yamamoto T, Tada T et al. Endothelial dysfunction in ischemic acute renal failure: rescue by transplanted endothelial cells. Am J Physiol Renal Physiol 2002;282:F1140-9.

[43] Schnackenberg CG. Physiological and pathophysiological roles of oxygen radicals in the renal microvasculature. Am J Physiol Regul Integr Comp Physiol 2002;282:R335-42.

[44] Morisawa N, Sugano N, Yamakawa T, Kuriyama S, Yokoo T. Successful long-term effects of direct renin inhibitor aliskiren in a patient with atherosclerotic renovascular hypertension. CEN Case Rep 2017;6:66-73.

[45] Moriyama T, Tsuruta Y, Kojima C et al. Beneficial effect of aliskiren combined with olmesartan in reducing urinary protein excretion in patients with chronic kidney disease. Int Urol Nephrol 2012;44:841-5.

[46] Li SY, Chen YT, Yang WC et al. Effect of add-on direct renin inhibitor aliskiren in patients with non-diabetes related chronic kidney disease. BMC Nephrol 2012;13:89.

[47] Bryfogle JW, Bradley RF. The vascular complications of diabetes mellitus; a clinical study. Diabetes 1957;6:159-67.

[48] Courreges JP, Bacha J, Aboud E, Pradier P. Prevalence of renal artery stenosis in type 2 diabetes. Diabetes Metab 2000;26 Suppl 4:90-6.

[49] Fatica RA, Port FK, Young EW. Incidence trends and mortality in end-stage renal disease attributed to renovascular disease in the United States. Am J Kidney Dis 2001;37:1184-90.

[50] Mailloux LU, Bellucci AG, Mossey RT et al. Predictors of survival in patients undergoing dialysis. Am J Med 1988;84:855-62.

[51] Covic A, Gusbeth-Tatomir P. The role of the renin-angiotensin-aldosterone system in renal artery stenosis, renovascular hypertension, and ischemic nephropathy: diagnostic implications. Prog Cardiovasc Dis 2009;52:204-8.

[52] Safian RD, Textor SC. Renal-artery stenosis. N Engl J Med 2001;344:431-42.

[53] Postma CT, van Oijen AH, Barentsz JO et al. The value of tests predicting renovascular hypertension in patients with renal artery stenosis treated by angioplasty. Arch Intern Med 1991;151:1531-5.

[54] Gloviczki ML, Glockner JF, Lerman LO et al. Preserved oxygenation despite reduced blood flow in poststenotic kidneys in human atherosclerotic renal artery stenosis. Hypertension 2010;55:961-6.

[55] Chen K, Xie F, Liu S et al. Plasma reactive carbonyl species: Potential risk factor for hypertension. Free Radic Res 2011;45:568-74.

[56] Wang W, Saad A, Herrmann SM et al. Changes in inflammatory biomarkers after renal revascularization in atherosclerotic renal artery stenosis. Nephrol Dial Transplant 2016;31:1437-43.

[57] Kinra M, Mudgal J, Arora D, Nampoothiri M. An insight into the role of cyclooxygenase and lipooxygenase pathway in renal ischemia. Eur Rev Med Pharmacol Sci 2017;21:5017-5020.

[58] Dias AT, Rodrigues BP, Porto ML et al. Sildenafil ameliorates oxidative stress and DNA damage in the stenotic kidneys in mice with renovascular hypertension. J Transl Med 2014;12:35.

[59] Zhang X, Eirin A, Li ZL et al. Angiotensin receptor blockade has protective effects on the poststenotic porcine kidney. Kidney Int 2013;84:767-75.

[60] Reckelhoff JF, Romero DG, Yanes Cardozo LL. Sex, Oxidative Stress, and Hypertension: Insights From Animal Models. Physiology (Bethesda) 2019;34:178-188.

[61] Grisk O, Rettig R. Interactions between the sympathetic nervous system and the kidneys in arterial hypertension. Cardiovasc Res 2004;61:238-46.

[62] DiBona GF, Kopp UC. Neural control of renal function. Physiol Rev 1997;77:75-197.

[63] Osborn JW, Foss JD. Renal Nerves and Long-Term Control of Arterial Pressure. Compr Physiol 2017;7:263-320.

[64] Azizi M, Sapoval M, Gosse P et al. Optimum and stepped care standardised antihypertensive treatment with or without renal denervation for resistant hypertension (DENERHTN): a multicentre, open-label, randomised controlled trial. Lancet 2015;385:1957-65.

[65] Krum H, Schlaich M, Whitbourn R et al. Catheter-based renal sympathetic denervation for resistant hypertension: a multicentre safety and proof-of-principle cohort study. Lancet 2009;373:1275-81.

[66] Krum H, Schlaich MP, Sobotka PA et al. Percutaneous renal denervation in patients with treatment-resistant hypertension: final 3-year report of the Symplicity HTN-1 study. Lancet 2014;383:622-9.

[67] Townsend RR, Mahfoud F, Kandzari DE et al. Catheter-based renal denervation in patients with uncontrolled hypertension in the absence of antihypertensive medications (SPYRAL HTN-OFF MED): a randomised, sham-controlled, proof-of-concept trial. Lancet 2017;390:2160-2170.

[68] Osborn JW, Banek CT. Catheter-Based Renal Nerve Ablation as a Novel Hypertension Therapy: Lost, and Then Found, in Translation. Hypertension 2018;71:383-388.

[69] Mahfoud F, Cremers B, Janker J et al. Renal hemodynamics and renal function after catheter-based renal sympathetic denervation in patients with resistant hypertension. Hypertension 2012;60:419-24.

[70] Guyenet PG. The sympathetic control of blood pressure. Nat Rev Neurosci 2006;7:335-46.

[71] Peterson JR, Sharma RV, Davisson RL. Reactive oxygen species in

the neuropathogenesis of hypertension. Curr Hypertens Rep 2006;8:232-41.

[72] Brody MJ. Central nervous system and mechanisms of hypertension. Clin Physiol Biochem 1988;6:230-9.

[73] Gordon FJ, Haywood JR, Brody MJ, Johnson AK. Effect of lesions of the anteroventral third ventricle (AV3V) on the development of hypertension in spontaneously hypertensive rats. Hypertension 1982;4:387-93.

[74] Whyte DG, Johnson AK. Thermoregulatory role of periventricular tissue surrounding the anteroventral third ventricle (AV3V) during acute heat stress in the rat. Clin Exp Pharmacol Physiol 2005;32:457-61.

[75] Loperena R, Harrison DG. Oxidative Stress and Hypertensive Diseases. Med Clin North Am 2017;101:169-193.

[76] Zimmerman MC, Lazartigues E, Lang JA et al. Superoxide mediates the actions of angiotensin II in the central nervous system. Circ Res 2002;91:1038-45.

[77] Lob HE, Schultz D, Marvar PJ, Davisson RL, Harrison DG. Role of the NADPH oxidases in the subfornical organ in angiotensin II-induced hypertension. Hypertension 2013;61:382-7.

[78] Nagashima Y, Ohaki Y, Tanaka Y et al. Establishment of an epithelioid malignant schwannoma cell line (YST-1). Virchows Arch B Cell Pathol Incl Mol Pathol 1990;59:321-7.

[79] Zimmerman MC, Lazartigues E, Sharma RV, Davisson RL. Hypertension caused by angiotensin II infusion involves increased superoxide production in the central nervous system. Circ Res 2004;95:210-6.

[80] Zimmerman MC, Sharma RV, Davisson RL. Superoxide mediates angiotensin II-induced influx of extracellular calcium in neural cells. Hypertension 2005;45:717-23.

[81] Grassi G, Seravalle G, Brambilla G, Mancia G. The sympathetic nervous system and new nonpharmacologic approaches to treating hypertension: a focus on renal denervation. Can J Cardiol 2012;28:311-7.

[82] Wang D, Jose P, Wilcox CS. beta(1) Receptors protect the renal afferent arteriole of angiotensin-infused rabbits from norepinephrine-induced oxidative stress. J Am Soc Nephrol 2006;17:3347-54.

[83] Boivin V, Jahns R, Gambaryan S, Ness W, Boege F, Lohse MJ. Immunofluorescent imaging of beta 1- and beta 2-adrenergic receptors in rat kidney. Kidney Int 2001;59:515-31.

[84] Campese VM, Ye S. A vitamin-E-fortified diet reduces oxidative stress, sympathetic nerve activity, and hypertension in the phenol-renal injury model in rats. J Am Soc Hypertens 2007;1:242-50.

[85] DiBona GF, Esler M. Translational medicine: the antihypertensive effect of renal denervation. 2010;298:R245-53.

[86] Chade AR, Krier JD, Rodriguez-Porcel M et al. Comparison of acute and chronic antioxidant interventions in experimental renovascular disease. Am J Physiol Renal Physiol 2004;286:F1079-86.

[87] Chade AR, Rodriguez-Porcel M, Herrmann J et al. Antioxidant intervention blunts renal injury in experimental renovascular disease. J Am Soc Nephrol 2004;15:958-66.

[88] Chade AR, Rodriguez-Porcel M, Grande JP et al. Distinct renal injury in early atherosclerosis and renovascular disease. Circulation 2002;106:1165-71.

[89] Chade AR, Rodriguez-Porcel M, Grande JP et al. Mechanisms of renal

structural alterations in combined hypercholesterolemia and renal artery stenosis. Arterioscler Thromb Vasc Biol 2003;23:1295-301.

[90] Mazzolai L, Duchosal MA, Korber M et al. Endogenous angiotensin II induces atherosclerotic plaque vulnerability and elicits a Th1 response in ApoE-/- mice. Hypertension 2004;44:277-82.

[91] Welch WJ, Mendonca M, Aslam S, Wilcox CS. Roles of oxidative stress and AT1 receptors in renal hemodynamics and oxygenation in the postclipped 2K,1C kidney. Hypertension 2003;41:692-6.

[92] Cuevas S, Zhang Y, Yang Y et al. Role of renal DJ-1 in the pathogenesis of hypertension associated with increased reactive oxygen species production. Hypertension 2012;59:446-52.

[93] Liu R, Ren Y, Garvin JL, Carretero OA. Superoxide enhances tubuloglomerular feedback by constricting the afferent arteriole. Kidney Int 2004;66:268-74.

[94] Wang D, Chen Y, Chabrashvili T et al. Role of oxidative stress in endothelial dysfunction and enhanced responses to angiotensin II of afferent arterioles from rabbits infused with angiotensin II. J Am Soc Nephrol 2003;14:2783-9.

[95] Araujo M, Wilcox CS. Oxidative stress in hypertension: role of the kidney. Antioxid Redox Signal 2014;20:74-101.

[96] Chabrashvili T, Kitiyakara C, Blau J et al. Effects of ANG II type 1 and 2 receptors on oxidative stress, renal NADPH oxidase, and SOD expression. 2003;285:R117-24.

[97] Meng S, Cason GW, Gannon AW, Racusen LC, Manning RD, Jr. Oxidative stress in Dahl salt-sensitive hypertension. Hypertension 2003;41:1346-52.

[98] Nagase M, Shibata S, Yoshida S, Nagase T, Gotoda T, Fujita T. Podocyte injury underlies the glomerulopathy of Dahl salt-hypertensive rats and is reversed by aldosterone blocker. Hypertension 2006;47:1084-93.

[99] Raij L, Tian R, Wong JS, He JC, Campbell KN. Podocyte injury: the role of proteinuria, urinary plasminogen, and oxidative stress. Am J Physiol Renal Physiol 2016;311:F1308-F1317.

[100] Jia J, Ding G, Zhu J et al. Angiotensin II infusion induces nephrin expression changes and podocyte apoptosis. Am J Nephrol 2008;28:500-7.

[101] Hua P, Feng W, Rezonzew G, Chumley P, Jaimes EA. The transcription factor ETS-1 regulates angiotensin II-stimulated fibronectin production in mesangial cells. Am J Physiol Renal Physiol 2012;302:F1418-29.

[102] Ren Z, Liang W, Chen C, Yang H, Singhal PC, Ding G. Angiotensin II induces nephrin dephosphorylation and podocyte injury: role of caveolin-1. Cell Signal 2012;24:443-50.

[103] Beswick RA, Dorrance AM, Leite R, Webb RC. NADH/NADPH oxidase and enhanced superoxide production in the mineralocorticoid hypertensive rat. Hypertension 2001;38:1107-11.

[104] Cai H, Li Z, Dikalov S et al. NAD(P)H oxidase-derived hydrogen peroxide mediates endothelial nitric oxide production in response to angiotensin II. J Biol Chem 2002;277: 48311-7.

[105] Landmesser U, Cai H, Dikalov S et al. Role of p47(phox) in vascular oxidative stress and hypertension caused by angiotensin II. Hypertension 2002;40:511-5.

[106] Suzuki H, Swei A, Zweifach BW, Schmid-Schonbein GW. In vivo

evidence for microvascular oxidative stress in spontaneously hypertensive rats. Hydroethidine microfluorography. Hypertension 1995;25:1083-9.

[107] Swei A, Lacy F, DeLano FA, Schmid-Schonbein GW. Oxidative stress in the Dahl hypertensive rat. Hypertension 1997;30:1628-33.

[108] Zhou X, Bohlen HG, Miller SJ, Unthank JL. NAD(P)H oxidase-derived peroxide mediates elevated basal and impaired flow-induced NO production in SHR mesenteric arteries in vivo. 2008;295:H1008-H1016.

[109] Lerman L, Textor SC. Pathophysiology of ischemic nephropathy. Urol Clin North Am 2001;28:793-803, ix.

[110] Napoli C, de Nigris F, Palinski W. Multiple role of reactive oxygen species in the arterial wall. J Cell Biochem 2001;82:674-82.

[111] Rodriguez-Porcel M, Lerman A, Herrmann J et al. Hypertension exacerbates the effect of hypercholesterolemia on the myocardial microvasculature. Cardiovasc Res 2003;58:213-21.

[112] Rodriguez-Porcel M, Krier JD, Lerman A et al. Combination of hypercholesterolemia and hypertension augments renal function abnormalities. Hypertension 2001;37:774-80.

[113] Rodriguez-Porcel M, Herrman J, Chade AR et al. Long-term antioxidant intervention improves myocardial microvascular function in experimental hypertension. Hypertension 2004;43:493-8.

[114] Benna JE, Dang PM, Gaudry M et al. Phosphorylation of the respiratory burst oxidase subunit p67(phox) during human neutrophil activation. Regulation by protein kinase C-dependent and independent pathways. J Biol Chem 1997;272:17204-8.

[115] Bouin AP, Grandvaux N, Vignais PV, Fuchs A. p40(phox) is phosphorylated on threonine 154 and serine 315 during activation of the phagocyte NADPH oxidase. Implication of a protein kinase c-type kinase in the phosphorylation process. J Biol Chem 1998;273:30097-103.

[116] el Benna J, Faust LP, Babior BM. The phosphorylation of the respiratory burst oxidase component p47phox during neutrophil activation. Phosphorylation of sites recognized by protein kinase C and by proline-directed kinases. J Biol Chem 1994;269:23431-6.

[117] El Benna J, Faust RP, Johnson JL, Babior BM. Phosphorylation of the respiratory burst oxidase subunit p47phox as determined by two-dimensional phosphopeptide mapping. Phosphorylation by protein kinase C, protein kinase A, and a mitogen-activated protein kinase. J Biol Chem 1996;271:6374-8.

[118] Faust LR, el Benna J, Babior BM, Chanock SJ. The phosphorylation targets of p47phox, a subunit of the respiratory burst oxidase. Functions of the individual target serines as evaluated by site-directed mutagenesis. J Clin Invest 1995;96:1499-505.

[119] Li H, Wallerath T, Munzel T, Forstermann U. Regulation of endothelial-type NO synthase expression in pathophysiology and in response to drugs. Nitric Oxide 2002;7:149-64.

[120] Hink U, Li H, Mollnau H et al. Mechanisms underlying endothelial dysfunction in diabetes mellitus. Circ Res 2001;88:E14-22.

[121] Mollnau H, Wendt M, Szocs K et al. Effects of angiotensin II infusion on the

expression and function of NAD(P)H oxidase and components of nitric oxide/cGMP signaling. Circ Res 2002; 90:E58-65.

[122] Munzel T, Li H, Mollnau H et al. Effects of long-term nitroglycerin treatment on endothelial nitric oxide synthase (NOS III) gene expression, NOS III-mediated superoxide production, and vascular NO bioavailability. Circ Res 2000;86:E7-E12.

[123] Zou X, Kwon SH, Jiang K et al. Renal scattered tubular-like cells confer protective effects in the stenotic murine kidney mediated by release of extracellular vesicles. Sci Rep 2018;8:1263.

[124] Cavalcanti CO, Alves RR, de Oliveira AL et al. Inhibition of PDE5 Restores Depressed Baroreflex Sensitivity in Renovascular Hypertensive Rats. Front Physiol 2016;7:15.

[125] Nguyen G, Delarue F, Burckle C, Bouzhir L, Giller T, Sraer JD. Pivotal role of the renin/prorenin receptor in angiotensin II production and cellular responses to renin. J Clin Invest 2002;109:1417-27.

[126] Batenburg WW, Krop M, Garrelds IM et al. Prorenin is the endogenous agonist of the (pro)renin receptor. Binding kinetics of renin and prorenin in rat vascular smooth muscle cells overexpressing the human (pro)renin receptor. J Hypertens 2007;25:2441-53.

[127] Nabi AH, Kageshima A, Uddin MN, Nakagawa T, Park EY, Suzuki F. Binding properties of rat prorenin and renin to the recombinant rat renin/prorenin receptor prepared by a baculovirus expression system. Int J Mol Med 2006;18:483-8.

[128] Nurun NA, Uddin NM, Nakagawa T et al. Role of "handle" region of prorenin prosegment in the non-proteolytic activation of prorenin by binding to membrane anchored (pro) renin receptor. Front Biosci 2007;12:4810-7.

[129] Zhang J, Noble NA, Border WA, Owens RT, Huang Y. Receptor-dependent prorenin activation and induction of PAI-1 expression in vascular smooth muscle cells. Am J Physiol Endocrinol Metab 2008;295:E810-9.

[130] Burckle CA, Jan Danser AH, Muller DN et al. Elevated blood pressure and heart rate in human renin receptor transgenic rats. Hypertension 2006;47:552-6.

[131] Ichihara A, Hayashi M, Kaneshiro Y et al. Inhibition of diabetic nephropathy by a decoy peptide corresponding to the "handle" region for nonproteolytic activation of prorenin. J Clin Invest 2004;114:1128-35.

[132] Ichihara A, Sakoda M, Kurauchi-Mito A, Kaneshiro Y, Itoh H. Involvement of (pro)renin receptor in the glomerular filtration barrier. J Mol Med (Berl) 2008;86:629-35.

[133] Ichihara A, Sakoda M, Kurauchi-Mito A, Kaneshiro Y, Itoh H. Renin, prorenin and the kidney: a new chapter in an old saga. J Nephrol 2009;22:306-11.

[134] Ichihara A, Sakoda M, Mito-Kurauchi A, Itoh H. Activated prorenin as a therapeutic target for diabetic nephropathy. Diabetes Res Clin Pract 2008;82 Suppl 1:S63-6.

[135] Kaneshiro Y, Ichihara A, Sakoda M et al. Slowly progressive, angiotensin II-independent glomerulosclerosis in human (pro)renin receptor-transgenic rats. J Am Soc Nephrol 2007;18:1789-95.

[136] Nabi AH, Suzuki F. Biochemical properties of renin and prorenin binding to the (pro)renin receptor. Hypertens Res 2010;33:91-7.

Chapter 5

The Role of Renin Angiotensin Aldosterone System in the Progression of Cognitive Dysfunction in Chronic Kidney Disease Patients with Alzheimer's Disease

Vinothkumar Ganesan

Abstract

Renin angiotensin aldosterone (RAAS) is very well established as a regulator of blood pressure (BP) and a determinant of target organ injury. It controls fluid and electrolyte balance through coordinated effects on the heart, blood vessels, and kidneys. The main effector of RAAS is angiotensin II (Ang II), which exerts its vasoconstrictor effect primarily on the postglomerular arterioles, thereby raising the glomerular hydraulic pressure and ultrafiltration of plasma proteins, which may lead to the initiation and progression of chronic kidney disease (CKD). RAAS also plays a, key role in hypertension and cerebrovascular disease. Enhanced Ang II levels accelerate the initiation and progression of cell senescence by fostering inflammation and oxidative stress. Sustained activation of RAAS facilitates aging-related CKD and results in cognitive dysfunction and Alzheimer's disease (AD). However, in many hypertension treatment studies, the frequency of fatal and nonfatal stroke has been greatly reduced, and this is very important since a history of stroke doubles the risk of dementia in both patients without CKD and hemodialysis. In CKD patients with AD, anemia has also been identified as a risk factor for cognitive impairment, and correction of anemia with recombinant erythropoietin treatment has been shown to enhance cognition measures, such as AD markers and neuropsychological tests.

Keywords: Angiotensin converting enzyme, Chronic Kidney Disease, Cognitive Dysfunction, Alzheimer's disease, Amyloid β, Tau

1. Introduction

The renin angiotensin aldosterone (RAAS) system is a hormone system in the body that is responsible for controlling the balance of fluid and blood pressure. The system consists primarily of three hormones, namely renin, angiotensin II and aldosterone. It is controlled mainly by the rate of renal blood flow. The main effector of RAAS is angiotensin II (Ang II), Rising glomerular hydraulic pressure and ultra-filtration of plasma proteins, which can contribute to the initiation and progression of chronic kidney disease (CKD), as well as key molecules in hypertension and cerebrovascular disease, exerts its vasoconstrictor effect primarily on postglomerular arterioles. Enhanced Ang II levels speed up the initiation and progression of cell senescence by encouraging inflammation and oxidative stress. Sustained RAAS activation facilitates aging-related CKD and results in cognitive decline and Alzheimer's disease (AD). The risk of cognitive dysfunction in CKD patients with AD is significantly greater than in patients without CKD [1], not only in aged patients with CKD, but also in young patients with CKD [2]. It has been believed for a long time that kidney function is associated with brain activity. Our recent clinical studies indicate that CKD patients are more vulnerable to cognitive dysfunction and AD, and the severity of cognitive dysfunction is closely linked to the development of CKD and kidney failure [3–5].

2. RAAS: pathogenic mechanism of chronic kidney disease

RAAS is the best known blood pressure regulator (BP) and the determinant of hypertension damage to the target organs. It also regulates the balance of fluids and electrolytes by coordinated impacts on the heart, blood vessels, and kidneys. The main effector of the RAAS is Ang II [6]. Renin is secreted from the juxtaglomerular apparatus of the kidney in the classic RAAS pathway and acts on the circulating precursor angiotensinogen to create angiotensin I. Angiotensin I has few effects on BP, and in the lungs, ACE is transformed to Ang II. Ang II operates on the heart and kidneys by binding to type 1 (AT1) and type 2 (AT2) G-protein coupled receptors [7]. The more deleterious effects of Ang II, vasoconstriction and heart and vessel hypertrophy are mediated by the AT1 receptor. In addition the vasodilator peptide bradykininin is inactivated by the angiotensin-converting enzyme (ACE)

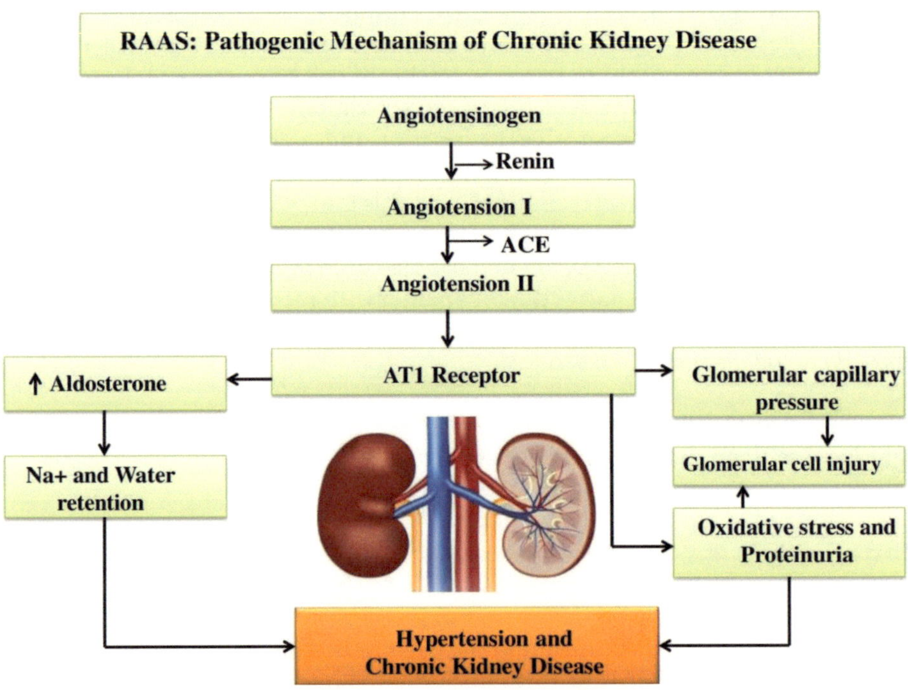

Figure 1.
The pathogenic mechanism of chronic kidney disease in the renin angiotensin aldosterone system.

in addition to the conversion of angiotensin I to Ang II [7]. Recently, ACE type 2 (ACE2) has been found to cleave angiotensin I into inactive angiotensin1–9, transformed by ACE into vasodilator and antiproliferative angiotensin1–7, respectively [8, 9]. While ACE2 in the human kidney is known to be present, there was no evidence on the distribution of tissues in kidney disease [8]. Kidney biopsies from patients with different kidney disorders, including transplant patients, were studied in a recent review. ACE2 was present in the tubular and glomerular epithelium and in the vascular smooth muscle cells and the interlobular artery endothelium in the control kidneys [10]. Neo-expression of ACE2 has been observed in the glomerular and peritubular capillary endothelium in all kidney diseases. Treatment with ACE inhibitors did not change ACE2 expression [10]. In vivo, Ang II increases the vascular tone of both afferent and efferent glomerular arterioles and modulates capillary intraglomerular pressure and glomerular filtration rate (GFR). Ang II primarily exerts its vasoconstrictor effect on the postglomerular arterioles, thereby raising the glomerular hydraulic pressure and filtration fraction, and impairing the glomerular barrier's selective size role for macromolecules, such as plasma proteins [11]. Intra capillary hypertension and increased plasma protein ultrafiltration can lead to the onset and progression of CKD [12]. Angiotensin non-hemodynamic effect may also be relevant in the progression of kidney disease [6].

A diagrammatic sketch of the pathogenic role of RAAS in chronic kidney disease is shown in **Figure 1**.

3. RAAS: pathogenic mechanism of Alzheimer's disease

Alzheimer disease (AD) is the most common neurodegenerative disease associated with dementia in the elderly. Various mechanisms, including DNA damage, lysosomal dysfunction, epigenetic modulation, and immune dysregulation, have been involved in neurodegenerative pathogenesis. Importantly, the homeostasis between protein synthesis, folding, and clearance of unfolded proteins, called proteostasis, is disrupted in AD and other neurodegenerative diseases. This contributes to an accumulation of proteins that are oligomerized and aggregated (Intracellular Tau (Neurofibrillary tangles [NFT]), and extracellular amyloid β (Aβ) (Senile plaques)) that ultimately induce protein toxicity. In many neurodegenerative disorders, including AD, oxidative stress are frequently found. In AD, Aβ accumulation, tau hyperphosphorylation, and the resulting degradation of synapses and neurons may be promoted by oxidative stress. In several target cells, Ang II has been shown to induce mitochondrial dysfunction through angiotensin II type 1 receptor (AT1R). Mechanistically, Ang II increases mitochondrial reactive oxygen species (ROS) [13]. Several studies indicate that ROS is involved in the development of Aβ fibrillation and NFT in AD and increases the pathology of Aβ and NFT in AD [14, 15].

The hyperactivity of the RAAS classical axis, mediated by AT1R, is implicated in the pathogenesis of AD. Ang II intracerebroventricular infusion increased both of the amyloid-β (Aβ) [16] and tau pathology, and also reduced cognitive performance [17], in aged normal rats. In most but not all AD mouse models, angiotensin II type 1 receptor blockers (ARBs) and angiotensin-converting enzyme inhibitors (ACEIs) minimize the amount of AD-like pathology and increase cognitive efficiency [18, 19]. Clinical studies have also identified ACE2 and ACE as brain RAAS factors, not only in the regulation of blood pressure, but also in the conversion of Aβ43 to Aβ40, which may decrease Aβ accumulation associated with AD and decrease serum ACE-2 activity in AD patients compared to control subjects [20].

A diagrammatic sketch of the pathogenic role of RAAS in Alzheimer's disease is shown in **Figure 2**.

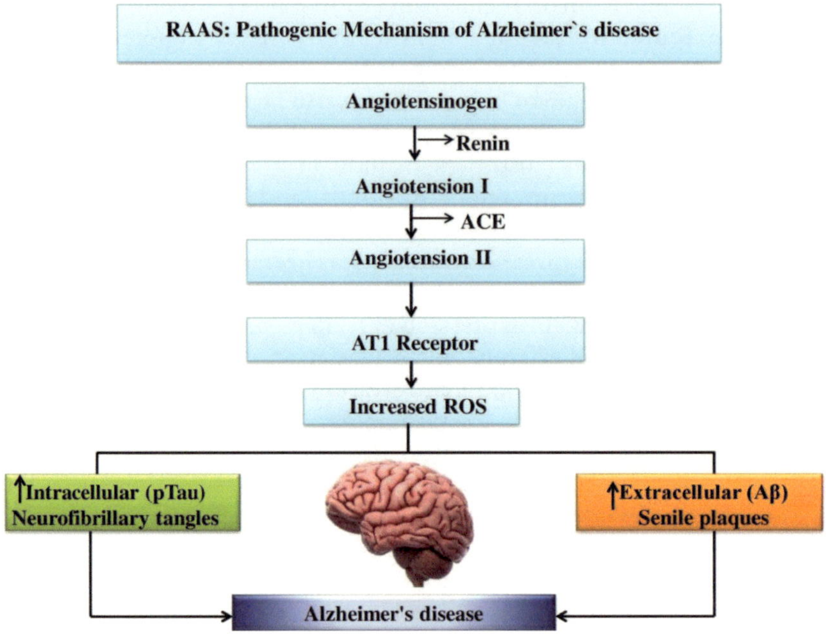

Figure 2.
Pathogenic Alzheimer's disease pathway of the renin angiotensin aldosterone system.

4. Hypertension is a risk factor for cognitive dysfunction in chronic kidney disease patients with Alzheimer's disease

The most common neurodegenerative disorders associated with CKD in the elderly are AD and dementia. Ang II represents a central molecule in cerebrovascular pathology and hypertension. Enhanced Ang II levels speed up the initiation and progression of cell senescence by encouraging inflammation and oxidative stress. Sustained RAAS activation causes aging related end stage organ damage and results in cognitive decline and dementia [21]. Studies also show that hypertension is the most important factor that adversely affects cerebral aging modalities and is related to cognitive compromise in people who are aging [22, 23]. This discovery has contributed to the belief that hypertension, up to the point of AD and dementia, is one of the factors responsible for the compromise of cognitive function in the elderly. It is therefore hypothesized that aging contributes to systemic and tissue RAAS hyperactivity and a rise in neurogenic hypertension, whereas evidence that connects brain RAAS with AD, memory, and learning develops cognitive functions [24]. In this regard, one of the long-term hypertension complications is clinically defined as dementia (for example AD) or vascular dementia, associated with diseases of the degenerative central nervous system (CNS). The temporal association between dementia and broad cerebrovascular pathology indicates that there is a pattern of sudden initiation and progressive development of cognitive impairment in the onset of dementia within three months of the diagnosis of stroke. It is understood that hypertension raises the risks of the target organ, such as cardiomegaly, progressive hypertensive retinopathy, nephropathy and stroke. In addition to repeated episodes of stroke or acute ischemic attacks, chronic hypertension, which results in a reduction in cerebral blood flow, is associated with vascular dementia and results in cognitive impairment [25].

A diagrammatic sketch of the role of RAAS in the induction and mediation of high blood pressure and cognitive impairment in CKD patients with AD is shown in **Figure 3**.

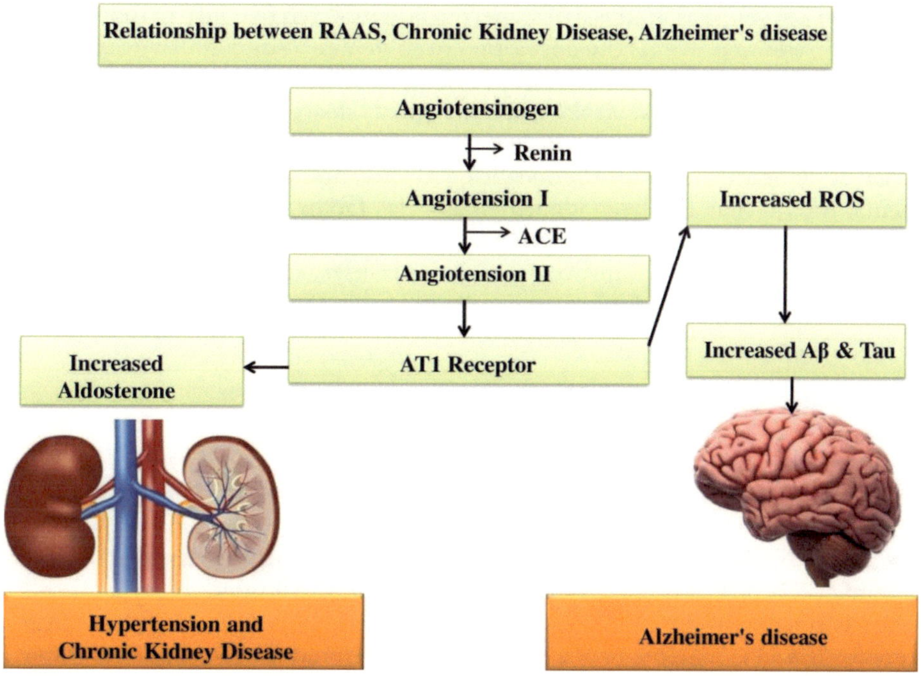

Figure 3.
Chronic kidney disease and alzheimer's associated renin angiotensin aldosterone system share ageing related molecular pathways, including processing of APP, tau phosphorylation, and increased oxidative stress.

5. Treatment of cognitive dysfunction in chronic kidney disease patients with Alzheimer's disease

Cognitive dysfunction is common among patients with CKD and dialysis in the memory, attention, and executive function domains. In our previous study, working memory and executive control, two main areas of cognitive ability, are potentially significant variables in drug compliance [4, 5]. Increased risk for injury, increased health care costs and progression to dementia are also associated with cognitive dysfunction without dementia [26]. Dementia is described by a drop in cognitive performance from a previous higher level along with a behavioral disorder that interferes with everyday function and independence. The brain and kidney vascular beds have identical anatomical and hemodynamic characteristics; these results have contributed to the speculation that cognitive dysfunction and CKD are a reflection of vascular damage in multiple end organs [26]. In addition, most patients with CKD have elevated rates of hypertension, diabetes, high levels of inflammatory receptors and vascular endothelial dysfunction, cardiovascular events like stroke, and carotid atherosclerosis, both leading to vascular cognitive decline and neurodegenerative diseases such as AD [27]. Potential steps to minimize cognitive impairment in CKD patients may include the treatment of cardiovascular risk factors, but, unfortunately, no clinical trials have been performed in CKD patients assessing cardiovascular risk factors for the prevention of cerebrovascular disease or cognitive impairment.

There is a trial showing that treatment with hypertension has a beneficial effect on cognition. In that survey, High blood pressure care with medication not only improves the cardiovascular health of older people, but can also reduce their risk of dementia and AD [27, 28]. The combined risk ratio of dementia preferred care in a meta-analysis of antihypertensive trials [29]. There is no strong evidence from the trials in a systematic analysis of hypertension research to confirm that decreasing blood pressure prevents the development of dementia or cognitive decline in hypertensive patients with no clear previous CVD [30]. However, the occurrence of fatal and non-fatal stroke has been greatly decreased in many studies of hypertension treatment, and it is very important since a history of stroke doubles the risk of dementia in both patients with non CKD and *hemodialysis*. In CKD patients, Anemia has also been identified as a risk factor for cognitive decline in CKD patients, and our studies have shown correction of anemia with recombinant erythropoietin therapy to improve cognitive measures, such as AD markers and neuropsychological tests [4, 5].

6. Future directions and challenges

This chapter explores the relationship between RAAS, cognitive dysfunction anemic CKD patients and EPO. We then hypothesized that the EPO may inhibit ACE2 interest and likely eventually alter complicated signaling cascades to boost cognition through changes in AD markers. A main aspect of this assessment is that in anemic CKD sufferers with cognitive impairment, the limited molecular effects of the treatment with EPO are crystal-clear. I may conclude by saying that a bright future for the EPO remedy. In order to better understand the mechanisms underlying the effects of EPO in anemic CKD with AD patients, further research into pharmacogenomics and clinical trials is required.

Author details

Vinothkumar Ganesan
Department of Medical Research, SRM Medical College Hospital and Research Centre, SRM Institute of Science and Technology, Kattankulathur, India

*Address all correspondence to: vinothkumarbiochem@gmail.com

References

[1] Ito H, Antoku S, Mori T, Nakagawa Y, Mizoguchi K, Matsumoto S, Omoto T, Shinozaki M, Nishio S, Abe M, Mifune M. Association between chronic kidney disease and the cognitive function in subjects without overt dementia. Clinical Nephrology. 2018; 1;89(5):330.

[2] Hailpern SM, Melamed ML, Cohen HW, Hostetter TH. Moderate chronic kidney disease and cognitive function in adults 20 to 59 years of age: Third National Health and Nutrition Examination Survey (NHANES III). Journal of the American Society of Nephrology. 2007; 1;18 (7):2205-13.

[3] Vinothkumar G, Kedharnath C, Krishnakumar S, Sreedhar S, Preethikrishnan K, Dinesh S, Sundaram A, Balakrishnan D, Shivashekar G, Sureshkumar VP. Abnormal amyloid β42 expression and increased oxidative stress in plasma of CKD patients with cognitive dysfunction: a small scale case control study comparison with Alzheimer's disease. BBA Clin.2017; 8:20-27

[4] Vinothkumar G, Krishnakumar S, Shivashekar G, Sreedhar S, Dinesh S, Sundaram A, Balakrishnan D, Riya VP. Therapeutic impact of rHuEPO on abnormal platelet APP, BACE 1, presenilin 1, ADAM 10 and Aß expressions in chronic kidney disease patients with cognitive dysfunction like Alzheimer's disease: a pilot study. Biomed Pharmacother. 2018; 104:211-222

[5] Vinothkumar G, Krishnakumar S, Riya VP. Correlation between abnormal GSK3ß, ß amyloid, total tau, p-tau 181 levels and neuro psychological assessment total scores in CKD patients with cognitivdysfunction: impact of rHuEPO therapy. J Clin Neurosci. 2019; 69:38-42.

[6] Aros C, Remuzzi G: The renin-angiotensin system in progression, remission and regression of chronic nephropathies. J Hyperten 20 (Suppl 3):S45–S53, 2002

[7] Zaman MA, Oparil S, Calhoun DA: Drugs targeting the renin–angiotensin–aldosterone system. Nat Rev 1:621-636, 2002

[8] Boehm M, Nabel EG: Angiotensin-converting enzyme 2—A new cardiac regulator. N Engl J Med 347:1795-1797, 2002

[9] Donoghue M, Hsieh F, Baronas E, et al: A novel angiotensinconverting enzyme-related carboxypeptidase (ACE2) converts angiotensin I to angiotensin. 2000; 1-9. Circ Res 87:E1–E9, Lely AT, Hamming I, Van Goor H, et al: Renal ACE2 expression in human kidney disease. J Pathol. 2004; 204:587-593

[10] Lely AT, Hamming I, Van Goor H, et al: Renal ACE2 expression in human kidney disease. J Pathol.2004; 204:587-593

[11] Yoshioka T, Rennke HG, Salant DJ, et al: Role of abnormally high transmural pressure in the permselectivity defect of glomerular capillary wall:Astudy in early passive Heymann nephritis. Circ Res.1987; 61:531-538

[12] Remuzzi G, Bertani T. Pathophysiology of progressive nephropathies. N Engl J Med.1998; 339:1448-1456

[13] Forrester SJ, Booz GW, Sigmund CD, Coffman TM, Kawai T, Rizzo V, Scalia R, Eguchi S. Angiotensin II signal transduction: an update on mechanisms of physiology and pathophysiology. Physiol Rev. 2018;98:1627-1738.

[14] Martínez E., Navarro A., Ordóñez C., del Valle E., Tolivia J. Oxidative stress induces apolipoprotein D overexpression in hippocampus during aging and alzheimer's disease. Journal of Alzheimer's Disease. 2013;36(1):129-144.

[15] Zhao Y., Zhao B. Oxidative stress and the pathogenesis of alzheimer's disease. Oxidative Medicine and Cellular Longevity. 2013;2013:10.

[16] Zhu D, Shi J, Zhang Y, Wang B, Liu W, Chen Z, et al. Central angiotensin II stimulation promotes β amyloid production in Sprague Dawley rats. PLoS One. 2011;6:e16037.

[17] Tian M, Zhu D, Xie W, Shi J. Central angiotensin II-induced Alzheimer-like tau phosphorylation in normal rat brains. FEBS Lett. 2012;586:3737-3745.

[18] Dong YF, Kataoka K, Tokutomi Y, Nako H, Nakamura T, Toyama K, et al. Perindopril, a centrally active angiotensin-converting enzyme inhibitor, prevents cognitive impairment in mouse models of Alzheimer's disease. FASEB J. 2011;25:2911-2920.

[19] Ongali B, Nicolakakis N, Tong XK, Aboulkassim T, Papadopoulos P, Rosa-Neto P, et al. Angiotensin II type 1 receptor blocker losartan prevents and rescues cerebrovascular, neuropathological and cognitive deficits in an Alzheimer's disease model. Neurobiol Dis. 2014;68:126-136.

[20] Liu S, Liu J, Miura Y, Tanabe C, Maeda T, Terayama Y, et al. Conversion of Aβ43 to Aβ40 by the successive action of angiotensin-converting enzyme 2 and angiotensin-converting enzyme. J Neurosci Res. 2014;92:1178-1186.

[21] Bodiga VL, Bodiga S. Renin angiotensin system in cognitive function and dementia. Asian journal of Neuroscience. 2013

[22] Phillips S and Whisnant J. "Hypertension and stroke," in Hypertension: Pathophysiology, Diagnosis, and Management, J. Laragh and B. Brenner, Eds, Raven Press, New York, NY, USA, 2nd edition, 1990; 417-431

[23] Strandgaard S, Paulson OB. Cerebrovascular consequences of hypertension. The Lancet. 1994 Aug 20;344(8921):519-521

[24] Phillips MI and De Oliveira EM. "Brain renin angiotensin in disease," Journal of Molecular Medicine, 2008;86; 6;715-722

[25] Andel R, Hughes TF, and Crowe M, "Strategies to reduce the risk of cognitive decline and dementia," Aging Health, 2005;1, 107-116

[26] Tamura MK, Yaffe K. Dementia and cognitive impairment in ESRD: diagnostic and therapeutic strategies. Kidney international. 2011;1;79(1):14-22.

[27] Langa KM, Foster NL, Larson EB. Mixed dementia: emerging concepts and therapeutic implications. 2004; 15; 292(23):2901-8.

[28] Williamson, JD, Miller, ME, Bryan, RN, Lazar, RM, Coker, LH, Johnson, J, Cukierman, T, Horowitz, KR, Murray, A, Launer, LJ. "The Action to Control Cardiovascular Risk in Diabetes Memory in Diabetes Study (ACCORD-MIND): rationale, design, and methods". Am J Cardiol. 2007; 99 112i-122i

[29] Ding J, Davis-Plourde KL, Sedaghat S, Tully PJ, Wang W, Phillips C, Pase MP, Himali JJ, Windham BG, Griswold M, Gottesman R. Antihypertensive medications and risk for incident dementia and Alzheimer's disease: a meta-analysis of individual participant data from prospective

cohort studies. The Lancet Neurology. 2020; 1;19(1):61-70

[30] Sofi, F, Valecchi, D, Bacci, D, Abbate, R, Gensini, GF, Casini, A, Macchi, C. "Physical activity and risk of cognitive decline: a meta-analysis of prospective studies". J Intern Med. 2011; 269,107-117.

Chapter 6

Renin Angiotensin Aldosterone System, Glucose Homeostasis, and Prevention of Type 2 Diabetes: Mechanistic Insights and Evidence from Major Clinical Trials

Samara Skwiersky, Sandra Iwuala, Seeta Chillumuntala, Deborah Osafehinti and Jocelyne Karam

Abstract

With its alarmingly rising prevalence worldwide, type 2 diabetes has become a leading cause of morbidity and mortality around the planet. Efforts to prevent progression to diabetes in individuals at risk could have a significant positive public health impact. Multiple trials examining cardiovascular outcomes of Renin-Angiotensin-Aldosterone System (RAAS) inhibitors revealed, in secondary analysis, a significantly reduced risk of new onset diabetes in participants receiving these agents. This glycemic protective effect is attributed to the known implication of RAAS in the development of insulin resistance and type 2 diabetes. The DREAM trial and the NAVIGATOR trial were two large randomized controlled studies examining, as primary outcome, the effect of Ramipril and Valsartan respectively on the incidence of diabetes in patients with prediabetes. Their results confirmed a favorable glycemic effect of RAAS inhibition agents and suggested a possible added benefit of diabetes prevention to their other several cardiovascular and blood pressure benefits.

Keywords: diabetes prevention, renin-angiotensin- aldosterone system, glucose homeostasis, ACE inhibitors, angiotensin receptor blockers, prediabetes

1. Introduction

Diabetes Mellitus (DM) is a chronic disease characterized by hyperglycemia due to impaired glucose regulation [1]. Glucose regulation is controlled by insulin, a protein hormone produced and secreted by the β-cells of the pancreas. Type 2 diabetes mellitus (T2DM) is characterized by insulin resistance and impaired β-cell function, eventually leading to decreased insulin secretion.

Prediabetes is the disease state which precedes the diagnosis of diabetes [2]. It is characterized by hyperglycemia caused by insulin resistance and β-cell dysfunction, as is type 2 diabetes, but before serum glucose levels reach that of diabetic diagnostic thresholds. Just as in diabetes, the diagnosis of prediabetes is made based on results of fasting plasma glucose, oral glucose tolerance test, Hemoglobin A1c, and/or random serum glucose levels [3]. Prediabetes can be defined by impaired fasting glucose (IFG) with a fasting plasma glucose level 100–126 mg/dL (5.5–7.0 mmol/L), impaired glucose tolerance (IGT) with glucose level of 140–200 mg/dL (7.8–11.1 mmol/L) at 2 hours of the oral glucose tolerance test (OGTT), and/or HbA1c level of 5.7–6.5% (39–48 mmol/mol) [2, 3].

As the prevalence of diabetes continues to increase, it has become a severe public health problem worldwide. According to the World Health Organization (WHO) and International Diabetes Foundation (IDF), 451 million adults were diagnosed with diabetes worldwide in 2017, which was drastically increased from 108 million in 1980 [2–4]. This number is expected to increase to 693 million by the year 2045 [3]. According to the CDC, in 2015, approximately half (48.3%) of the adult population ages 65 and older had prediabetes [2].

With its many microvascular and macrovascular complications, diabetes contributes to a large portion of healthcare costs worldwide. In fact, approximately 850 billion USD of the global healthcare expenditure was spent on patients with diabetes in 2017 [2]. Research has shown that individuals with diabetes are at increased risk of cardiovascular disease (CVD), the leading cause of death worldwide [1]. The Framingham Heart study found that women with diabetes had a five times greater risk of heart failure, while men had two times greater risk, when compared to individuals of the same age and gender without diabetes [5]. Prediabetes has also been found to be independently associated with microvascular complications, macrovascular complications (including CVD) and increased risk of overall mortality [6, 7].

Aside from the increased risk of CVD in individuals with diabetes, an independent association between hypertension and insulin resistance has been established [8]. The Hong Kong Cardiovascular Risk Factor Prevalence Study found that of individuals with diabetes, 58% had elevated blood pressure, and of people with hypertension, 34% had impaired glucose tolerance. Only 42% of subjects studied with diabetes had normal blood pressure [9]. While the mechanism of this relationship is unclear, it has been hypothesized that patients with hypertension have impaired glucose tolerance due to changes in skeletal muscle tissue [10]. This common coexistence of hypertension and diabetes increases one's risk of CVD and events, and thus contributes to the increased risk of morbidity and mortality in these patients. Both hypertension and insulin resistance are components of the cardiometabolic syndrome, a group of interrelated abnormalities, which increase the risk for CVD and T2DM. Other related abnormalities include obesity, left ventricular hypertrophy, dyslipidemia, and albuminuria [10, 11].

Given the increasing prevalence of diabetes worldwide and its many complications, a significant effort has been made to explore preventive modalities. Studies have concluded that lifestyle interventions involving diet and physical activity reduce the risk of diabetes by greater than 50% [12]. However, the intense lifestyle modifications necessary to result in change are often difficult to implement. Bariatric surgery has been found to be an effective method of diabetes prevention and treatment. In a meta-analysis of 22,094 patients who had undergone bariatric surgery, diabetes was completely resolved in 76.8% of patients [13]. The Swedish Obese Subject Study, a prospective study of 4047 patients without diabetes who underwent gastric surgery, found that after 15 years, 392 of 1658 control patients developed diabetes compared to 110 of 1771 patients who underwent bariatric surgery ($p < 0.001$) [14]. Pharmacological agents such as metformin, thiazolidinediones, alpha-glucosidase inhibitors, and the glucagon-like peptide-1 agonist, liraglutide have been shown to prevent diabetes in those at risk [1, 15]. However, none of these agents have the added benefit of hypertension or CVD prevention and/or treatment. In fact, thiazolidinediones have been associated with an increased risk of congestive heart failure [12].

Pharmacological agents which act by inhibition of the Renin-Angiotensin-Aldosterone System (RAAS) including Angiotensin-converting enzyme inhibitors

DOI: http://dx.doi.org/10.5772/intechopen.97737

Trial	Year	Study Population (No.)	Drug of interest	Comparison	Diabetes prevention (primary or secondary)	Results
DREAM	2006	IGT and/or IGF (5269)	Ramipril	Placebo	Primary	Non-significant decrease in new-onset DM (HR 0.91, 95% CI 0.80–1.03), Significantly increased regression of IFG and IGT to normoglycemia, decrease in OGTT 2 hr. glucose level.
NAVIGATOR	2008	IGT + > 50 y/o with CVD OR > 55 y/o with > 1 RF for CVD (9518)	Valsartan	Placebo	Primary	significantly reduced DM incidence by 14% (HR 0.86, 95% CI 0.80–0.92)
HOPE	2001	Multiple cardiovascular risk factors (9297)	Ramipril	Placebo	Secondary	significantly reduced risk of new-onset DM (RR 0.66, 95% CI 0.41–0.77)
CAPP	1999	Ages 25–66 with HTN (DBP > 100 mmHg) (10,985)	Captopril	diuretics and/or B-blocker	Secondary	significantly decreased risk of new-onset DM with captopril by 14% (RR 0.86; 95% CI, 0.74–0.99)
ALLHAT	2002	Hypertension + 1 or more CHD Risk factor (MI, stroke, LVH, T2DM, smoking, low HDL, other atherosclerotic CVD) (42,418)	Lisinopril	Chlorthalidone, amlodipine	Secondary	significantly decreased risk of new-onset DM among patients taking lisinopril (8.1%) vs. amlodipine (9.8%) and chlorthalidone (11.6%)
LIFE	2002	Ages 55–80, hypertension + left ventricular hypertrophy (9193)	Losartan	Atenolol	Secondary	significantly decreased risk of new-onset DM in patients taking losartan compared with atenolol (HR 0.75; 95% CI, 0.63–0.88; p < 0.001)
PEACE	2004	> 50 years, coronary heart disease with LVEF > 40% (8.290)	Trandolapril	Placebo	Secondary	significantly decreased incidence of new onset DM in patients taking Trandolapril (HR 0.83; 95% CI, 0.72–0.96; p=.01)
VALUE	2004	HTN and high CV risk (male, >50 years, DM, current smoker, high TC, LVH, proteinuria) (15,245)	Valsartan	Amlodipine	Secondary	significantly decreased incidence of new-onset DM in patients taking Valsartan compared with Amlodipine (HR 0.77; 95% CI 0.69–0.86; p<.0001)

Table 1.
Trials with diabetes prevention as a primary and secondary outcome of RAAS inhibition.

(ACE-I) and angiotensin receptor blockers (ARBs) have been observed to have a favorable glycemic effect, and are among candidates examined in recent diabetes prevention trials. While they are often utilized for their blood pressure-lowering effect, they have cardiovascular benefits as well. Specifically, ACE-I have been found to play a role in the reversal of left ventricular hypertrophy in patients with hypertension, and preventing left ventricular remodeling post myocardial infarction [16]. Thus, ACE-I are indicated as first line agents in patients with heart failure, left ventricular systolic dysfunction (LVEF < 40–45%) and those with acute coronary syndrome and after suffering from an acute myocardial infarction [16]. In patients with heart failure, ACE-I have been shown to reduce mortality, hospitalizations, and prevent worsening of heart failure in these individuals [16]. The benefits of ARBs are less well defined, however, the clinical trial Val-HeFT found treatment with ARB, valsartan, resulted in decreased morbidity and mortality in patients with heart failure, when compared with placebo [17]. Additionally, ARBs have been found to slow the progression of diabetic nephropathy thus preventing end stage renal disease (ESRD) in these patients. Two trials, IDNT and RENAAL conducted in 2001, revealed ARBs (Irbesartan and Losartan) to be effective in reducing proteinuria and slowing the progression of ESRD in patients with diabetic nephropathy, independent of their blood-pressure lowering effect [18, 19].

Given these benefits, RAAS inhibitors are often first line agents for treating patients with concomitant hypertension and diabetes and those at risk for CVD. Several studies to date suggest that ACE-I and ARBs have the ability to improve glycemic control by improving insulin sensitivity. **Table 1** provides a brief description of the studies and their findings. This chapter explores the possibility of utilizing RAAS inhibitors as a means of diabetes prevention and/or improved glucose tolerance and the potential mechanisms by which this could be accomplished.

2. RAAS and glucose homeostasis

The renin-angiotensin-aldosterone system (RAAS) is responsible for regulating arterial blood pressure and blood volume [20, 21]. Renin, an enzyme produced by the juxtaglomerular cells in the kidney in response to low blood pressure or decreased sodium delivery to the kidneys, converts angiotensinogen to angiotensin I. Angiotensin converting enzyme (ACE), found in the lungs and kidneys, then converts angiotensin I to angiotensin II (AG II). Angiotensin II is the predominant hormone responsible for the hemodynamic effects of RAAS, namely: sodium retention at the proximal convoluted tubules of the kidneys, arterial vasoconstriction, and release of aldosterone from the adrenal glands [22]. Angiotensin II is also responsible for the non-hemodynamic effect of RAAS related to glucose hemostasis [21, 23]. Several studies have suggested the role of RAAS in the development of insulin resistance and subsequent development of type 2 diabetes mellitus (T2DM) in humans. The pathophysiology is complex, mostly involving the skeletal muscle, adipose tissue, and pancreas [21] (**Figure 1**).

1. **RAAS and the skeletal muscle:** AG II affects glucose metabolism in the skeletal muscle through the inhibition of insulin-mediated glucose uptake and insulin signaling pathway, and a decrease in the blood supply to the skeletal muscle [21].

   Inhibition of insulin-mediated glucose uptake and insulin signaling pathway. The skeletal muscle accounts for up to 70% of insulin-mediated glucose uptake in the body, which occurs through a series of tightly regulated events in the insulin signaling pathway [23, 24]. First, insulin binds to the insulin receptor on the surface of the skeletal muscle cell, and this activates a cascade of events that ultimately ends in translocation of the glucose transporters (GLUT-4) from intracellular vesicles to the cell membrane through which glucose is taken up by the cells [23, 24]. Therefore, inhibition at any stage in the signaling pathway will result in insulin resistance with subsequent type 2 diabetes development if left unresolved. By acting through the angiotensin II type 1 receptor (AT_1R), AG II activates NADPH oxidase, which leads to the production of reactive oxygen species that in turn inhibits insulin-mediated translocation of GLUT-4 transporters, glucose uptake, and insulin signaling pathway in the skeletal muscle [23, 24].

Decrease in the blood supply to the skeletal muscle. Studies also show that AG II contributes to insulin resistance by decreasing microvascular blood supply to the skeletal muscle [21].

2. **RAAS and the adipose tissue:** Studies have shown that local RAAS present in adipose tissue affects adipocyte differentiation through angiotensin II's action on its AT_1R receptor [21], but there are conflicting views on the exact mechanism. For example, some studies suggest that AG II inhibits adipocyte precursor differentiation, thereby decreasing the number of insulin-sensitive adipocytes leading to insulin resistance [25]. In contrast, other studies indicate that AG II stimulates adipocyte differentiation and causes an increase in adipocyte size in visceral adipose tissue leading to obesity and insulin resistance [21].

3. **RAAS and the pancreas:** Increased activity of local pancreatic RAAS is associated with impaired glucose metabolism. By acting through the AT_1R receptor, AG II decreases insulin secretion, impairs blood flow to the pancreatic islet cells, and causes inflammation and fibrosis of the pancreas, leading to impaired glucose tolerance [21].

In summary, through its different effects on the skeletal muscle, adipose tissue, and pancreas, RAAS is thought to contribute to the development of insulin resistance and development of type 2 diabetes. Therapy with RAAS inhibitors has been

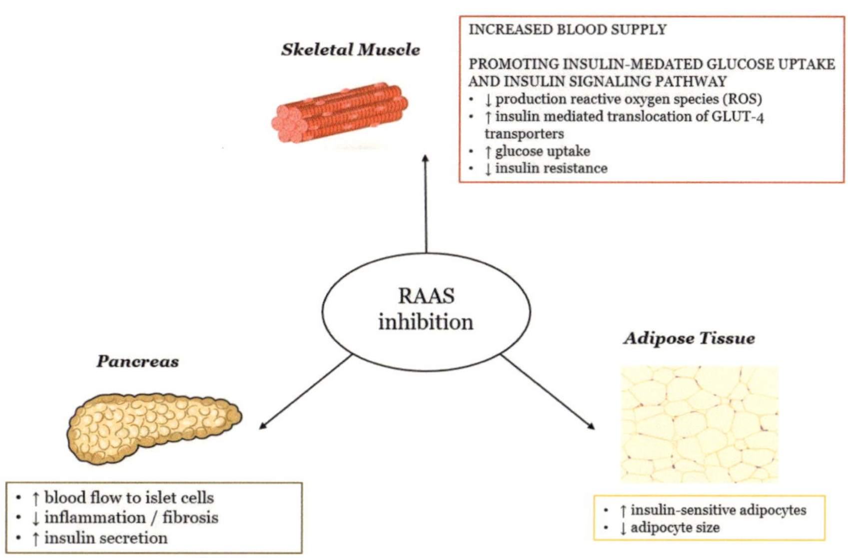

Figure 1.
Potential mechanisms implicated in favorable glycemic effect associated with RAAS inhibition.

indeed associated with favorable glycemic events. At a clinical level, several trials have examined the role of RAAS inhibition in preventing the development of type 2 diabetes in the population at risk.

3. Diabetes prevention as a secondary outcome of RAAS inhibition trials

There have been a number of trials conducted in which the primary aim was to study the effect of RAAS inhibitors on CVD and events. In addition to this primary outcome of interest, a number of these trials have found positive results with regards to their effect on diabetes prevention and improved glucose tolerance.

One of the first clinical trials to demonstrate a protective effect of RAAS inhibition on the incidence of diabetes was the Captopril Prevention Project (CAPPP) initiated in 1999. The primary aim of this trial was to compare the effect of ACE inhibition (using captopril) with conventional therapy (β-blockers and/or diuretics) on risk of CVD morbidity and mortality in patients with hypertension [26]. While there was no difference in prevention of cardiovascular morbidity and mortality in those treated with captopril compared with conventional therapy, authors did find that the incidence of new onset diabetes was lower in participants treated with captopril [26]. This finding supports the theory that ACE inhibition may work to prevent the development of diabetes, which may be due to captopril's ability to improve insulin sensitivity [26]. Additionally, those patients with diabetes at baseline who were treated with captopril had a lower rate of cardiovascular events and mortality when compared to those with diabetes treated with conventional therapy [26].

Another study, the Heart Outcomes Prevention Evaluation (HOPE) study, sought to explore the role of the ACE inhibitor, ramipril, on the incidence of myocardial infarction (MI), stroke, or all-cause mortality in patients with a history of vascular disease (coronary artery disease, stroke, peripheral vascular disease) or diabetes, plus at least one other cardiovascular risk factor (hypertension, elevated total cholesterol levels, low high-density lipoprotein cholesterol levels, cigarette smoking, or microalbuminuria), but without heart failure or any degree of left ventricular dysfunction [27]. Subjects were randomized to receive ramipril or placebo, both with the addition of 400 IU of vitamin E daily [27]. Of the primary outcomes examined, patients treated with ramipril had a significantly decreased risk of myocardial infarction, stroke, or death from cardiovascular causes (RR 0.78, 95% CI 0.70–0.86). Of the participants without a diagnosis of diabetes at study onset, there was a 34% decreased incidence of new onset diabetes in those treated with ramipril compared with placebo (RR 0.66, 95% CI 0.34–0.76) [27]. Of note, these results are consistent with the study to Evaluate Carotid Ultrasound changes in patients treated with ramipril and vitamin E (SECURE), which reported decreased fasting glucose levels in patients treated with ramipril when compared with placebo [28].

Another trial, the Losartan Intervention For Endpoint reduction in hypertension study (LIFE), randomized participants aged 55–80 years with hypertension and electrocardiographic left ventricular hypertrophy (ECG LVH) to either losartan or atenolol [29]. The primary aim of this trial was to determine whether losartan improves LVH and thus reduces cardiovascular morbidity and mortality. Results revealed that those participants who received losartan had a decreased risk of cardiovascular events (MI and stroke), and 25% decreased risk of new onset DM when compared with atenolol (HR 0.75, 95% CI 0.63–0.88, p 0.001) [30]. It is possible that the protective effect of losartan on diabetes incidence seen in the LIFE trial could be due to the detrimental effects of atenolol, a β-blocker, on insulin sensitivity [10].

The presence of diabetes has been found to be associated with increased left ventricular hypertrophy, both of which are risk factors for the cardiometabolic syndrome [29]. The initial analysis of the LIFE trial found that individuals treated with losartan had an increased regression of LVH when compared to those treated with atenolol. However, patients with diabetes and LVH had less regression than those without diabetes, possibly secondary to their predisposition [29]. A secondary analysis was conducted on the participants without diabetes at baseline, which sought to determine whether in-treatment resolution or continued absence of ECG LVH is associated with decreased risk of developing diabetes [29]. This analysis revealed a 38% decreased incidence of DM in those who had resolution or continued absence of LVH (HR 0.62, 95% CI 0.50–0.78, $p < 0.001$) independent of the previously identified effects of treatment with losartan versus atenolol. This finding suggests that while DM might lead to LVH, it is possible that LVH may in fact precede the development of diabetes [29]. While the causality of this relationship is uncertain, this study proposes the idea that regression of LVH by means of RAAS inhibition might decrease the risk for DM. However, it is also possible that this observed relationship between LVH regression and decreased incidence of DM can be explained by the established association between hypertension and insulin resistance. This idea aligns with the finding that participants of the LIFE trial who developed diabetes had higher baseline systolic and diastolic blood pressures than those who did not [29].

In the Antihypertensive and Lipid-Lowering Treatment to Prevent Heart Attack (ALLHAT) trial, the primary aim was to compare the effectiveness of treatment with diuretic, chlorthalidone against calcium channel blocker, amlodipine and ACE-I, lisinopril in preventing coronary heart disease (CHD) or other cardiovascular events in patients with hypertension and at least one CHD risk factor [31]. As far as primary outcome of interest, chlorthalidone was found to be superior to the others in preventing the primary outcome. However, study participants on lisinopril were found to have a lower incidence of diabetes at the follow up period of four years, when compared to those placed on other antihypertensives [31].

Similarly, the Prevention of Events with Angiotensin Converting Enzyme Inhibition (PEACE) trial sought to determine whether treatment with another ACE-I, trandolapril in patients with stable CAD and left ventricular ejection fraction (LVEF) > 40% would reduce cardiovascular deaths, incidence of MI or need for percutaneous coronary intervention (PCI) when compared with treatment with placebo [32]. Although a secondary end point, results from this trial revealed that the incidence of new onset DM was significantly decreased in those treated with trandolapril when compared to those in the placebo group (HR 0.83, 95%CI 0.72–0.96, $p=0.01$) [32]. Results from the PEACE trial, similar to the HOPE trial are important because they cannot be attributed to the adverse effects of the comparison drug (placebo).

The Valsartan Antihypertensive Long-Term Use Evaluation (VALUE) trial compared coronary heart disease outcomes in patients with hypertension treated with valsartan or amlodipine [33]. While there were no differences in primary composite outcome of cardiovascular morbidity and mortality in either group, secondary analysis revealed that new-onset DM occurred significantly less in patients treated with valsartan [33]. Despite the observed decreased incidence of diabetes with valsartan use, blood pressure reduction was less in this group, compared to those treated with amlodipine, which suggests that the effect of ARBs on diabetes prevention is independent of blood pressure reduction [34].

While each of the above trials found treatment with ACE-I and ARBs to be associated with decreased incidence of new onset diabetes, it must be noted that diabetes prevention was not a defined primary outcome in any of these studies.

Thus, their results must be interpreted with caution. A few other weaknesses should be taken into consideration on review. Some of these trials, including the HOPE and PEACE trials did not perform formal glucose testing to establish glycemic status, and relied on self-report alone [32, 35, 36]. Additionally, the HOPE, CAPP, and LIFE trials all utilized β-blockers as comparator drugs, which allows for the possibility that the observed effect of therapy with ACE-I or ARB on diabetes prevention is due to the detrimental effects of B-blockers on development of diabetes rather than the benefits of RAAS inhibition. A large prospective cohort study (n=12,550) conducted in 2000 revealed that hypertensive patients taking β-blockers had a 28% increased risk of diabetes when compared to those who were not on any antihypertensive therapy [37].

4. DREAM and NAVIGATOR trials

Studies including the aforementioned trials showed a beneficial effect of RAAS inhibition with ACE-I and ARBs on diabetes prevention among patients with hypertension and other cardiovascular diseases [30, 35, 38, 39]. These trials studied diabetes prevention as a secondary outcome or post hoc analysis, thus the results should be interpreted with caution. Conversely, the DREAM and NAVIGATOR trials, conducted in 2006 and 2008, respectively, are double blind, randomized clinical trials, which were designed to determine the effect of RAAS inhibition on the incidence of diabetes as a primary outcome [40, 41]. Furthermore, in these two trials, glycemic categories were meticulously determined, defined and recorded. In both studies, DM was defined using standard criteria, fasting blood glucose (FBG) 126 mg/dl or 200 mg/dl post oral glucose load and confirmed again at a later date. In the DREAM study, even in the event that diabetes was diagnosed by an outside physician, confirmation of the diagnosis using standard plasma glucose criteria was required in addition to the prescription of an antidiabetic agent by the diagnosing physician [40].

The DREAM trial was designed to investigate the effect of ramipril, an ACE-I and rosiglitazone, a thiazolidinedione, on diabetes prevention among patients with prediabetes (IGT and/or IFG) but without cardiovascular disease. The primary outcome of this study was newly diagnosed diabetes or death, with a secondary outcome of regression to normoglycemia defined as normal fasting and 2 hour post-load glucose levels [40]. Data analysis revealed no significant difference in the development of diabetes in the ramipril group when compared to the placebo group (HR 0.91, 95% CI 0.80–1.03) [40]. However, the likelihood of regression to normoglycemia was increased among subjects within the ramipril group when compared to the placebo group (HR 1.16, 95% CI 1.07–1.27). Moreover, while the fasting plasma glucose levels did not differ between the ramipril and the placebo group at the end of the trial, the 2 hour post glucose oral load values were significantly lower among those within the ramipril group [40].

There are a number of possible explanations for the lack of reduction in DM incidence with ramipril use in the DREAM trial which was different from the results found in previous trials with ACE-I/ARBs. First, as mentioned, diagnosis of diabetes at study onset was unambiguously established in participants of the DREAM trial with an oral glucose tolerance test (OGTT), thus patients with pre-existing DM were reliably excluded from the study [40]. This was not the case for some of the other studies mentioned previously [35, 42]. Second, the demographics of the DREAM study patients differed from those of trials which showed a reduced incidence of DM with RAAS inhibitors. Compared to the participants of the DREAM trial, subjects from the other trials were older, and had established CVD, and/or heart failure [30, 35, 36, 43, 44]. It is possible that the RAAS system is activated to a greater extent and thus ACE inhibition may have greater benefits in these individuals [45]. Third, some of the trials that revealed reduced incidence of DM among those treated with ACE-I/ARBs had compared ACE-I with other anti-hypertensives associated with dysglycemia, such as **β**-blockers, as mentioned previously. This may have led to a possible exaggeration of the effect of RAAS inhibition on diabetes prevention. Fourth, most of the previous trials that showed a beneficial effect of ACE-I And ARB on DM prevention followed the patients for longer period of time than the median 3 years that the participants of the DREAM trial were followed for [30, 32, 35, 39, 43]. Specifically, the participants of the HOPE trial were followed for 4.5 years, the PEACE trial for 4.8 years, ALLHAT study for 4.9 years, and the LIFE study for 4.8 years [30, 32, 35, 39]. In the DREAM trial, there was a late diversion of the Kaplan–Meier curves that suggested a benefit of ramipril in DM prevention after 3–5 years. Thus, it is possible that a longer and larger study may be needed to observe the effect of ramipril on DM prevention [45].

Four years after the publication of the results of the DREAM trial, the results of another trial, the Nateglinide and Valsartan in Impaired Glucose Tolerance Outcomes Research (NAVIGATOR) trial, were released [41]. This study also sought to investigate the effect of RAAS inhibition with the ARB, valsartan in addition to lifestyle modification on diabetes prevention in patients with impaired glucose tolerance and established CVD or CVD risk factors.

The NAVIGATOR trial was an improvement over the DREAM trial in several ways. First, a study-specific lifestyle-intervention program, which has previously been found to reduce risk of diabetes by up to 50%, was implemented for all patients in addition to pharmacotherapy [46, 47]. Second, there was a longer median follow up of 5 years in the NAVIGATOR trial compared with the 3 years follow up in the DREAM trial [40, 41]. Third, the NAVIGATOR trial enrolled a larger number of participants, 9306, versus 5269 participants in the DREAM trial. Another difference between these studies is that, unlike the DREAM trial, the NAVIGATOR trial enrolled patients with established CVD or CVD risk factors, who may have a greater degree of RAAS activation at baseline.

With these differences in mind, it is not surprising that while the DREAM trial found no difference between the ramipril and placebo groups with regards to diabetes prevention, the NAVIGATOR trial found that those treated with valsartan had a significantly reduced incidence of DM by 14% (HR 0.86, 95% CI 0.80–0.92, $p < 0.001$). Furthermore, patients in the valsartan arm of the study had lower mean fasting plasma glucose and 2 hours post glucose load levels. Additionally, the proportion of patients taking glucose lowering agents at the end of the study was lower in the valsartan group than in those in the placebo group.

Although significant, the 14% reduction in diabetes risk with valsartan appears smaller than the risk reduction seen in previously conducted trials involving ACE-I and ARBs [32, 35, 36, 44, 48]. One possible reason is that by the last study visit, a significantly higher proportion of subjects in the placebo arm were taking other ARBs or ACE-I (24.4% vs. 21.8%), which could have diluted the effect seen with valsartan. Another reason for this observed discrepancy could be due to a difference in the way in which glycemic status was determined at study onset and completion. Unlike the NAVIGATOR trial, a few other trials diagnosed DM by self-report rather than formal glucose testing which allows for misclassification error and possible false exaggeration of results [35].

In addition, the effect of valsartan with lifestyle modification was much smaller compared to landmark studies on diabetes prevention with lifestyle alone in which the incidence of DM was reduced by as much as 58% [46, 47, 49]. Similarly, the effect of valsartan on diabetes prevention in the NAVIGATOR trial is smaller when compared to glucose lowering agents such as metformin, 26–31% [46, 50], acarbose 25% [51] and rosiglitazone 60% in the DREAM study [52]. It is worthy of note that the NAVIGATOR trial followed the subjects for a longer duration (5 years) compared to the trials involving these glucose lowering agents in which subjects were followed for 2.5–3.3 years.

In conclusion, the DREAM and NAVIGATOR trials showed benefit in glycemic indices but only the NAVIGATOR trial showed a reduced diabetes incidence as a primary outcome of RAAS inhibition with ACE-I and ARBs. These findings may have utility in the clinical setting, in terms of choice of antihypertensive agents to those at higher risk of DM development, in the presence or absence of CVD and its risk factors.

5. Conclusion and clinical implications

ACE-I and ARBs are currently widely used for the treatment of patients with hypertension, heart failure or asymptomatic left ventricular dysfunction, coronary artery disease, and diabetic nephropathy, with the clinical benefits of ACE-I more closely studied [53]. Based on the results from the aforementioned trials, the use of these agents may also be indicated for the prevention of diabetes and/or regression from impaired to normoglycemia. This is extremely significant in light of the emerging diabetes epidemic.

While it is not entirely clear, results from the trials explored throughout this chapter suggest that those with cardiometabolic syndrome and its risk factors including (but not limited to) hypertension, obesity, insulin resistance, and left ventricular hypertrophy may experience the greatest benefits with regards to diabetes prevention and improved glycemic control. This could be due to the fact that the RAAS system is overactive in a number of these conditions. As discussed, activation of the RAAS system and increased production of angiotensin II is thought to play a role in the development of insulin resistance and subsequent development of T2DM [21]. It is also possible that the ability of ACE-I and ARBs to prevent diabetes is in part due to their effect on blood pressure reduction and LVH regression, both of which have been shown to improve insulin sensitivity [29].

However, while results from the CAPPP trial found a decreased incidence of new onset DM in patients treated with captopril, the blood pressure of patients in this group was significantly higher throughout the study than those treated with conventional therapy with β-blocker and/or diuretics. This supports the hypothesis that captopril's effect on diabetes prevention might be independent of blood pressure reduction. Results from the sub-analysis of the LIFE trial suggests that the effect of RAAS inhibition with losartan on LVH regression may be partly responsible for the decreased incidence of DM. It is possible that this association is also explained in part by the relationship between blood pressure and insulin resistance [29]. In conclusion, the apparent decreased incidence of new onset diabetes seen in patients treated with ACE-I and ARBs are likely attributable to both direct and indirect effects of these agents.

Given the variety of indications for which RAAS inhibitors have been established, the additional benefit of diabetes prevention could help to alleviate polypharmacy in individuals who suffer from several of these conditions simultaneously. However, more research is needed to categorically place ACE-I and ARBs among the armamentarium of agents favoring DM prevention. Head to head studies comparing the effects of different ACE-I and ARBs would also be useful.

Author details

Samara Skwiersky[1], Sandra Iwuala[1], Seeta Chillumuntala[1], Deborah Osafehinti[2] and Jocelyne Karam[1,2*]

1 SUNY-Downstate, Health Science University, Brooklyn, NY, USA

2 Maimonides Medical Center, Brooklyn, NY, USA

*Address all correspondence to: jkaram@maimonidesmed.org

References

[1] *Diabetes and the Eye: Latest Concepts and Practices*. Diabetes: Current and Future Developments. Vol. 2. 2021, Singapore: Bentham Science Publishers Pte. 249.

[2] Yip, W.C.Y., et al., *Prevalence of Pre-Diabetes across Ethnicities: A Review of Impaired Fasting Glucose (IFG) and Impaired Glucose Tolerance (IGT) for Classification of Dysglycaemia.* Nutrients, 2017. **9**(11): p. 1273.

[3] Cho, N.H., et al., *IDF Diabetes Atlas: Global estimates of diabetes prevalence for 2017 and projections for 2045.* Diabetes Research and Clinical Practice, 2018. **138**: p. 271-281.

[4] Organization, W.H. *Diabetes Fact Sheet*. 2020 [cited 2021 February 24, 2021]; Available from: https://www.who.int/news-room/fact-sheets/detail/diabetes.

[5] Kannel, W.B. and D.L. McGee, *Diabetes and cardiovascular disease. The Framingham study.* Jama, 1979. **241**(19): p. 2035-8.

[6] Karam JG., M.S., *Prevention of Type 2 Diabetes: Evidence and Strategies.* Journal of Clinical Outcomes Management, 2017.

[7] Huang, Y., et al., *Association between prediabetes and risk of cardiovascular disease and all cause mortality: systematic review and meta-analysis.* BMJ (Clinical research ed.), 2016. **355**: p. i5953-i5953.

[8] Bonora, E., et al., *Relationship between blood pressure and plasma insulin in non-obese and obese non-diabetic subjects.* Diabetologia, 1987. **30**(9): p. 719-723.

[9] Cheung, B.M., *The hypertension-diabetes continuum.* J Cardiovasc Pharmacol, 2010. **55**(4): p. 333-9.

[10] McFarlane, S.I., A. Kumar, and J.R. Sowers, *Mechanisms by which angiotensin-converting enzyme inhibitors prevent diabetes and cardiovascular disease.* The American journal of cardiology, 2003. **91**(12): p. 30-37.

[11] Castro, J.P., et al., *Cardiometabolic syndrome: pathophysiology and treatment.* Current hypertension reports, 2003. **5**(5): p. 393-401.

[12] Gerstein, H.C., et al., *Effect of rosiglitazone on the frequency of diabetes in patients with impaired glucose tolerance or impaired fasting glucose: a randomised controlled trial.* Lancet, 2006. **368**(9541): p. 1096-105.

[13] Buchwald, H., et al., *Bariatric surgery: a systematic review and meta-analysis.* Jama, 2004. **292**(14): p. 1724-37.

[14] Sjöström, C.D., *Surgery as an intervention for obesity. Results from the Swedish obese subjects study.* Growth Horm IGF Res, 2003. **13 Suppl A**: p. S22-6.

[15] *The Diabetes Prevention Program (DPP).* Description of lifestyle intervention, 2002. **25**(12): p. 2165-2171.

[16] Nasution, S.A., *The use of ACE inhibitor in cardiovascular disease.* Acta Med Indones, 2006. **38**(1): p. 60-4.

[17] Maggioni, A.P., et al., *Effects of valsartan on morbidity and mortality in patients with heart failure not receiving angiotensin-converting enzyme inhibitors.* J Am Coll Cardiol, 2002. **40**(8): p. 1414-21.

[18] Parving, H.H., et al., *The effect of irbesartan on the development of diabetic nephropathy in patients with type 2 diabetes.* N Engl J Med, 2001. **345**(12): p. 870-8.

[19] Keane, W.F. and P.A. Lyle, *Recent advances in management of type 2 diabetes and nephropathy: lessons from the RENAAL study.* Am J Kidney Dis, 2003. **41**(3 Suppl 1): p. S22-5.

[20] Atlas, S.A., *The renin-angiotensin aldosterone system: pathophysiological role and pharmacologic inhibition.* J Manag Care Pharm, 2007. **13**(8 Suppl B): p. 9-20.

[21] Underwood, P.C. and G.K. Adler, *The renin angiotensin aldosterone system and insulin resistance in humans.* Current hypertension reports, 2013. **15**(1): p. 59-70.

[22] Fountain, J.H. and S.L. Lappin, *Physiology, Renin Angiotensin System*, in *StatPearls*. 2020, StatPearls Publishing Copyright © 2020, StatPearls Publishing LLC.: Treasure Island (FL).

[23] Henriksen, E.J., *Improvement of insulin sensitivity by antagonism of the renin-angiotensin system.* Am J Physiol Regul Integr Comp Physiol, 2007. **293**(3): p. R974-80.

[24] McGuire, D.K., et al., *Blocking the renin-angiotensin-aldosterone system to prevent diabetes mellitus.* Diabetes and Vascular Disease Research, 2008. **5**(1): p. 59-66.

[25] Cooper, M.E., *The role of the renin-angiotensin-aldosterone system in diabetes and its vascular complications.* Am J Hypertens, 2004. **17**(11 Pt 2): p. 16S–20S; quiz A2-4.

[26] Hansson, L., et al., *Effect of angiotensin-converting-enzyme inhibition compared with conventional therapy on cardiovascular morbidity and mortality in hypertension: the Captopril Prevention Project (CAPPP) randomised trial.* Lancet, 1999. **353**(9153): p. 611-6.

[27] *Effects of an Angiotensin-Converting–Enzyme Inhibitor, Ramipril, on Cardiovascular Events in High-Risk Patients.* New England Journal of Medicine, 2000. **342**(3): p. 145-153.

[28] Lonn, E.M., et al., *Effects of Ramipril and Vitamin E on Atherosclerosis.* Circulation, 2001. **103**(7): p. 919-925.

[29] Okin, P.M., et al., *In-Treatment Resolution or Absence of Electrocardiographic Left Ventricular Hypertrophy Is Associated With Decreased Incidence of New-Onset Diabetes Mellitus in Hypertensive Patients.* Hypertension, 2007. **50**(5): p. 984-990.

[30] Dahlöf, B., et al., *Cardiovascular morbidity and mortality in the Losartan Intervention For Endpoint reduction in hypertension study (LIFE): a randomised trial against atenolol.* Lancet, 2002. **359**(9311): p. 995-1003.

[31] Officers, T.A. and C.f.t.A.C.R. Group, *Major Outcomes in High-Risk Hypertensive Patients Randomized to Angiotensin-Converting Enzyme Inhibitor or Calcium Channel Blocker vs DiureticThe Antihypertensive and Lipid-Lowering Treatment to Prevent Heart Attack Trial (ALLHAT).* JAMA, 2002. **288**(23): p. 2981-2997.

[32] Braunwald, E., et al., *Angiotensin-converting-enzyme inhibition in stable coronary artery disease.* N Engl J Med, 2004. **351**(20): p. 2058-68.

[33] Julius, S., et al., *VALUE trial: Long-term blood pressure trends in 13,449 patients with hypertension and high cardiovascular risk.* American journal of hypertension, 2003. **16**(7): p. 544-548.

[34] Bajuk Studen, K., M. Jensterle Sever, and M. Pfeifer, *Cardiovascular risk and subclinical cardiovascular disease in polycystic ovary syndrome.* Front Horm Res, 2013. **40**: p. 64-82.

[35] Yusuf, S., et al., *Ramipril and the development of diabetes.* Jama, 2001. **286**(15): p. 1882-5.

[36] Vermes, E., et al., *Enalapril Reduces the Incidence of Diabetes in Patients With Chronic Heart Failure.* Circulation, 2003. **107**(9): p. 1291-1296.

[37] Sowers, J.R. and G.L. Bakris, *Antihypertensive Therapy and the Risk of Type 2 Diabetes Mellitus.* New England Journal of Medicine, 2000. **342**(13): p. 969-970.

[38] Hansson, L., et al., *Effect of angiotensin-converting-enzyme inhibition compared with conventional therapy on cardiovascular morbidity and mortality in hypertension: the Captopril Prevention Project (CAPPP) randomised trial.* The Lancet, 1999. **353**(9153): p. 611-616.

[39] *Major outcomes in high-risk hypertensive patients randomized to angiotensin-converting enzyme inhibitor or calcium channel blocker vs diuretic: The Antihypertensive and Lipid-Lowering Treatment to Prevent Heart Attack Trial (ALLHAT).* Jama, 2002. **288**(23): p. 2981-97.

[40] Bosch, J., et al., *Effect of ramipril on the incidence of diabetes.* N Engl J Med, 2006. **355**(15): p. 1551-62.

[41] McMurray, J.J., et al., *Effect of valsartan on the incidence of diabetes and cardiovascular events.* N Engl J Med, 2010. **362**(16): p. 1477-90.

[42] Investigators, P.T., *Angiotensin-converting–enzyme inhibition in stable coronary artery disease.* New England Journal of Medicine, 2004. **351**(20): p. 2058-2068.

[43] Fox, K.M., *Efficacy of perindopril in reduction of cardiovascular events among patients with stable coronary artery disease: randomised, double-blind, placebo-controlled, multicentre trial (the EUROPA study).* Lancet, 2003. **362**(9386): p. 782-8.

[44] Yusuf, S., et al., *Effects of candesartan on the development of a new diagnosis of diabetes mellitus in patients with heart failure.* Circulation, 2005. **112**(1): p. 48-53.

[45] *Effect of Ramipril on the Incidence of Diabetes.* New England Journal of Medicine, 2006. **355**(15): p. 1551-1562.

[46] Knowler, W.C., et al., *Reduction in the incidence of type 2 diabetes with lifestyle intervention or metformin.* N Engl J Med, 2002. **346**(6): p. 393-403.

[47] Tuomilehto, J., et al., *Prevention of type 2 diabetes mellitus by changes in lifestyle among subjects with impaired glucose tolerance.* N Engl J Med, 2001. **344**(18): p. 1343-50.

[48] Kjeldsen, S.E., et al., *Effects of valsartan compared to amlodipine on preventing type 2 diabetes in high-risk hypertensive patients: the VALUE trial.* J Hypertens, 2006. **24**(7): p. 1405-12.

[49] Pan, X.R., et al., *Effects of diet and exercise in preventing NIDDM in people with impaired glucose tolerance. The Da Qing IGT and Diabetes Study.* Diabetes Care, 1997. **20**(4): p. 537-44.

[50] Ramachandran, A., et al., *The Indian Diabetes Prevention Programme shows that lifestyle modification and metformin prevent type 2 diabetes in Asian Indian subjects with impaired glucose tolerance (IDPP-1).* Diabetologia, 2006. **49**(2): p. 289-97.

[51] Chiasson, J.-L., et al., *Acarbose for prevention of type 2 diabetes mellitus: the STOP-NIDDM randomised trial.* The Lancet, 2002. **359**(9323): p. 2072-2077.

[52] Conlon, D., *DREAM — a diabetes prevention study.* The British Journal of Diabetes & Vascular Disease, 2006. **6**(5): p. 230-233.

[53] Sutter, J.D., M. Mendes, and O.H. Franco, *The ESC Textbook of Preventive Cardiology*, in *Cardioprotective drugs*. 2017, Oxford University Press.

Section 3

The Renin-Angiotensin Aldosterone System: Special Topics

Chapter 7

Diagnosis of Hypoaldosteronism in Infancy

Elpis-Athina Vlachopapadopoulou and Myrto Bonataki

Abstract

Hypoaldosteronism is associated with either insufficient aldosterone production or lack of responsiveness to aldosterone and can be isolated or in the context of primary adrenal failure. The severity of clinical manifestations is inversely correlated to age, with the neonatal period being the most vulnerable time for a patient to present with mineralocorticoid insufficiency. Salt-wasting forms of congenital adrenal hyperplasia (CAH), adrenal hypoplasia congenita (AHC), aldosterone synthase deficiency (ASD) and pseudohypoaldosteronism (PHA) are all causes of hypoaldosteronism in infancy. Affected infants present with salt wasting, failure to thrive and potentially fatal hyperkalemia and shock. A blood sample for the essential hormonal investigations should be collected before any steroid treatment is given, in order to confirm aldosterone insufficiency and to determine the underlying cause. Renal ultrasonography and urine culture are also useful for exclusion of secondary causes of aldosterone resistance. Initial management requires treatment of electrolyte imbalances and restoration of intravascular fluid volume. In case of a salt-wasting crisis, affected infants are usually treated initially with both hydrocortisone and fludrocortisone, pending the results of investigations. Interpretation of the hormonal profile will guide further therapy and molecular analysis of candidate genes.

Keywords: hypoaldosteronism, salt-wasting crisis, hyponatremia, hyperkalemia, pseudohypoaldosteronism

1. Introduction

Aldosterone, the most important mineralocorticoid, regulates electrolyte balance and intravascular volume by controlling renal sodium reabsorption and potassium excretion. Hypoaldosteronism is a rare, but potentially severe condition, associated with hyponatremia, hyperkalemia, metabolic acidosis and volume depletion. Given the higher mineralocorticoid demand during the critical neonatal period, the clinical presentation of aldosterone insufficiency in this age group can be dramatic [1–3].

2. The renin-angiotensin-aldosterone system in infancy

Regulation of fetal salt and water balance is handled by the placenta, so newborns with aldosterone defects have a normal electrolyte profile at birth.

Postnatally, healthy term neonates display a state of functional hypoaldosteronism (lower sodium and higher potassium concentrations), that contrasts with markedly increased aldosterone and renin secretion rates [2–5]. Indeed, it has been reported, that neonates have a mean plasma aldosterone level of 80 ng/dl versus 16.6 ng/dl for adults. Similarly, plasma renin activity (PRA) is severalfold higher in the first 3 months of life, than the levels reported later in adult life (450 and 25 ng $liter^{-1}$ min^{-1} respectively) [6]. Concurrent partial aldosterone resistance is attributed to the low mineralocorticoid receptor (MR) expression and the weak 11β-hydroxysteroid dehydrogenase type 2 (11βHSD2) activity of the neonatal kidney. Aldosterone unresponsiveness may, at least in part, account for the extracellular fluid compartment contraction and weight loss during the first days of life [7–9].

3. Clinical presentation of hypoaldosteronism

Biochemically, hypoaldosteronism is characterized by hyponatremia, hyperkalemia, prerenal azotemia and non-anion gap metabolic acidosis (hyperkalemic or type 4 renal tubular acidosis). The severity of clinical manifestations is inversely correlated to age, due to changes in adrenal and renal physiology [10, 11].

The critical neonatal period is clearly the most vulnerable time for an affected person to present with hypoaldosteronism [3]. Babies with mineralocorticoid insufficiency start to lose whole body sodium and water from day 1 in their urine. Overt electrolyte disturbances usually develop after the 4th day of life in infants with salt-wasting 21-hydroxylase deficiency, but a very early onset of symptoms may be seen in cases of severe systemic PHA. Affected infants eventually present with dehydration, vomiting and failure to thrive, while urine output remains excessively high for the degree of dehydration [6, 10]. Early warning signs, such as failure to reach birth weight by two weeks of age or excessive weight loss (greater than 10–12%) during the first days of life, should always prompt a careful assessment of hydration status, kidney function and electrolyte profile [7].

Clinical manifestations associated with hyponatremia are primarily neurologic, due to osmotic water shift intracellularly, parenchymal edema and brain ischemia. Frequent symptoms include vomiting, poor feeding and lethargy or irritability. Pronounced symptoms, such as seizures, are encountered rarely due to the insidious onset (>48 h) of hyponatremia. Open sutures and fontanelles in neonates, also act as a protective mechanism preventing intracranial hypertension [12, 13]. Yet infants with hypoaldosteronism are at risk of acute deterioration and might present in circulatory collapse with lethargy, tachycardia, hyperpnea, prolonged capillary refill, and cool and mottled extremities. Hypotension, a very late dehydration sign, occurs when all compensatory mechanisms to maintain organ perfusion have failed [14]. Hyperkalemia is clinically manifested by muscular weakness and cardiac disturbances (bradycardia, ventricular fibrillation, hypotension or cardiac arrest). Electrocardiogram (ECG) signs of hyperkalemia include repolarization abnormalities, peaked T-waves, QRS widening and depression of ST-segment. Arrhythmias may appear at any time and can lead to sudden death [15, 16].

Hypoaldosteronism has a much milder course in older children and adults, as aldosterone requirements normally decrease with age. Children with aldosterone insufficiency may present with subtle symptoms, such as postural hypotension and salt craving or even with asymptomatic growth failure. Autonomous addition of salt in the diet can delay or mask the presentation, until a simple viral gastroenteritis or a hot day associated with excessive sweating triggers the cascade of clinical manifestations [10, 11].

Reversible growth impairment is a well-known feature of several conditions accompanied by acidosis or electrolyte derangement (e.g., Bartter's syndrome or renal tubular acidosis) [6]. Accordingly, children with aldosterone insufficiency may present with linear growth deceleration due to chronic hyponatremia and acidosis. Sodium is an important growth factor, stimulating cell proliferation, protein synthesis and increasing cell mass. The mechanism whereby Na^+ promotes growth is through alkalinization of the cell interior, via a sodium-dependent Na^+/H^+-antiporter. Sodium depletion and acidosis lead to decreased antiporter system's activity and despite adequate macronutrient intake, children fail to thrive [17, 18].

4. Diagnostic workup

Hypoaldosteronism should be considered in any infant with persistent hyperkalemia and hyponatremia if there is no apparent cause, such as renal failure or prematurity. A diagnostic algorithm for infants presenting with salt-wasting and hyperkalemia (suspected mineralocorticoid defect) is presented in **Figure 1** [8, 16].

Renal function must be evaluated carefully, since renal excretion of potassium is directly dependent upon glomerular filtration rate (GFR). Renal adaptive mechanisms allow the kidneys to maintain potassium homeostasis until the GFR decreases to less than 15 ml/min/1.73 m^2 [16, 19].

Hyperkalemia and hyponatremia are well-recognized complications of prematurity. Hyperkalemia may be observed, even in the absence of oliguria, in very low birth weight preterm infants weighing less than 1,000 g. The serum potassium concentration may be as high as 9.0 mEq/1 and be accompanied by significant ECG irregularities [15]. Plasma potassium concentration decreases gradually from 6.5 ± 0.5 mEq/1 at 30— 32 weeks to 5.1 ± 0.2 mEq/1 at 39–41 weeks. Premature infants of <36 weeks gestational age (GA) are also unable to conserve sodium. The more immature the infant, the greater the risk and the degree of hyponatremia [20]. Urinary sodium excretion is 3.1 ± 0.5 mEq/kg/day (mean ± SE) in the newborn of 30–32 weeks gestational age and 1.2 ± 0.4 mEq/kg/day at 36—38 weeks gestational age [21].

The clinical scenario of a dehydrated infant with salt wasting and hyperkalemia represents a medical emergency implying inadequate mineralocorticoid action or complete adrenal insufficiency (both glucocorticoid and mineralocorticoid deficiency). The most common diagnosis in neonates is CAH due to 21-hydroxylase deficiency. Other conditions to be considered in infancy are the rare salt-wasting forms of CAH adrenal hypoplasia congenita, aldosterone synthase deficiency, PHA and drug effects [7, 22, 23].

Considerable overlap exists in the clinical and biochemical presentation of most of the above-mentioned endocrine diseases and only a few clinical signs can help clinicians to differentiate between them [23, 24]. A careful examination of the external genitalia is indicated in all infants with hyponatremia and hyperkalemia and might reveal valuable information towards the appropriate diagnosis [25]. 46, XX infants with 21-hydroxylase deficiency exhibit variable extent of virilization due to excessive androgen production. On the contrary, signs of undervirilization such as hypospadias are noted in 46,XY infants with 3β-hydroxysteroid dehydrogenase deficiency due to decreased androgen production. Infant boys with 21-hydroxylase deficiency have normal external genitalia or subtle penile enlargement that can be easily overlooked [25–29]. Impaired cortisol secretion is suggested clinically by low glucose levels and vascular tone insufficiency (hypotension) that is unresponsive to initial resuscitation. Last, although not always clinically obvious, increased pigmentation is a distinguishing

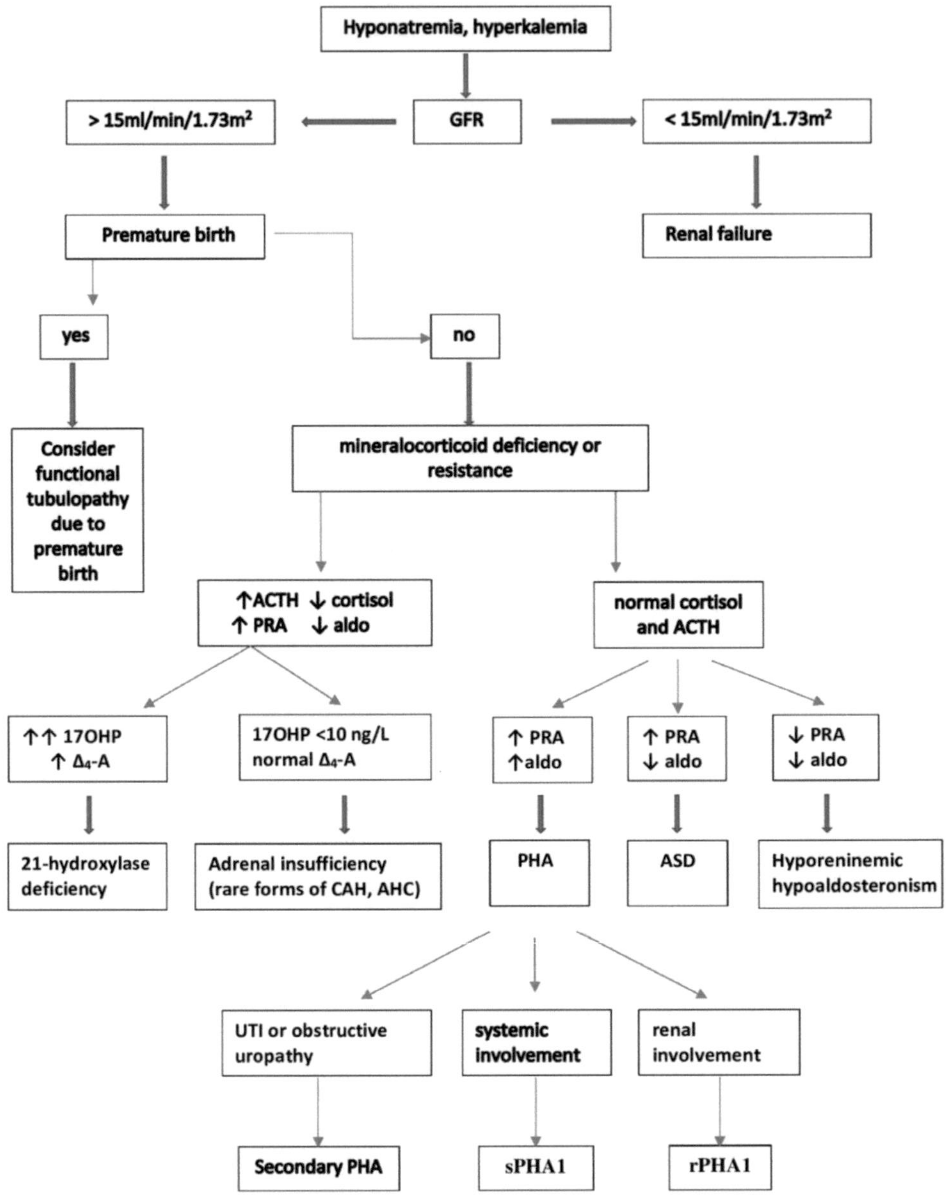

Figure 1.
Diagnostic approach to infants presenting with hyponatremia, hyperkalemia from references [8, 16] (GFR: glomerular filtration rate, PRA: plasma renin activity, 17OHP: 17-hydroxyprogesterone, Δ_4-A: Δ_4-androstenedione, Aldo: aldosterone, CAH: congenital adrenal hyperplasia, AHC: adrenal hypoplasia congenita, PHA: pseudohypoaldosteronism, sPHA: systemic PHA, rPHA: renal PHA, ASD: aldosterone synthase deficiency, UTI: urinary tract infection).

feature of primary adrenal insufficiency (PAI) associated with high levels of melanocyte-stimulating hormone (MSH), a ligand derived from pro-opiomelanocortin that causes hyperpigmentation of melanin-containing skin cells [6, 28].

The following investigations are suggested as a first - line diagnostic workup in infants with apparent mineralocorticoid deficiency:

- A critical sample for ACTH determination, cortisol, 17-hydroprogesterone (17- OHP), Δ_4 - androstenedione, aldosterone levels and renin level or PRA should be drawn before administration of hydrocortisone.

- A urine collection by suprapubic aspiration or catheterization for microscopic analysis, urine culture and urine electrolytes measurement.

- Abdominal ultrasonography [7, 10].

Hypoaldosteronism may be challenging to diagnose promptly, because aldosterone and renin assays are generally sent to a reference laboratory and results are usually delayed. Thus, urinary electrolyte assessment and abdominal ultrasonography are useful adjuncts in a clinical setting [12, 30].

4.1 Urine electrolytes

Urine sodium is expected to be low (usually <25) in hyponatremia, when the renal response is intact. A urine sodium concentration greater than 25 mEq/L demonstrates inappropriately high sodium excretion in the hyponatremic infant, suggesting aldosterone deficiency or resistance. Nevertheless, we should keep in mind that urinary sodium losses may not be excessive if the infant is salt depleted [12, 30, 31].

In addition to measuring urinary sodium, it is useful to estimate potassium excretion. The preferred method to estimate potassium excretion by the distal tubule is the transtubular potassium (K) concentration gradient (TTKG):

$$\mathrm{TTKG} = \left(\mathrm{Blood\ Osmolality} * \mathrm{Urine\ K}\right) / \left(\mathrm{Blood\ K} * \mathrm{Urine\ Osmolality}\right)$$

TTKG is expected to be high (>10) during hyperkalemia, as a result of appropriate aldosterone activity. TTKG values of less than six indicate impaired aldosterone action in the distal nephron as the cause of the hyperkalemia [31, 32]. Calculating TTKG before and after fludrocortisone administration is also useful in distinguishing patients who have mineralocorticoid deficiency versus resistance. An increase in TTKG values is observed after administering fludrocortisone in aldosterone-deficient states, but no change is seen in the case of aldosterone resistance [16].

4.2 Abdominal ultrasonography

Abdominal ultrasonography is a rapid, sensitive and non-invasive test that can provide valuable diagnostic information. The diagnosis of CAH is supported when a combination of 2 or more of the three following abnormalities are evident in adrenal sonography: (a) increased size (limb width > 4 mm), (b) lobulated or cerebriform surface and c) abnormal echogenicity. Adrenal imaging can also detect physical causes of adrenal insufficiency such as hemorrhage [33, 34].

Pelvic ultrasonography is indicated to evaluate internal genitourinary anatomy in infants with a suspected defect in steroidogenesis. Ultrasonography might reveal the presence of Müllerian structures in severely virilized infants with 21-hydroxylase deficiency carrying a 46,XX karyotype [35]. Accordingly, the lack of Müllerian structures in an infant with a salt-wasting crisis and female appearing external genitalia is found in 46,XY infants with classic congenital lipoid adrenal hyperplasia (CLAH) [36].

Urine culture and renal ultrasonography may allow early recognition of secondary PHA in infants with salt-wasting by detecting renal malformations and urinary tract infection (UTI). Further imaging with voiding cystourethrography (VCUG) and 99mTc-mercaptoacetyltriglycine (MAG3) scintigraphy may demonstrate vesicoureteral reflux (VUR) or obstruction [37].

4.3 Adrenal function tests

Evaluating apparent mineralocorticoid deficiency and eliciting the correct diagnosis requires a careful interpretation of adrenal function tests. Hypoaldosteronism can be isolated or in the context of PAI and concurrent cortisol production failure [38]. Serum cortisol level is expected to be elevated in the hypovolemic, acidotic patient with a functioning adrenal gland [39]. Diagnosis of PAI is suggested by an elevated plasma corticotropin (ACTH) concentration (frequently >100 pg/mL) in the presence of a low serum cortisol concentration (usually <10 mg/dL) [40]. Samples in young infants may be obtained at random, because diurnal secretion of ACTH and cortisol is not yet established, while by 6 months of age and beyond samples should be collected close to 8 AM [41]. In the setting of diagnostic uncertainty, confirmation of the diagnosis is established by an ACTH-stimulation test (250 μg for children >2 years, 15 μg/kg for infants and 125 μg for children <2 years, intravenously). A subnormal peak cortisol level (<18 μg/dL) 30 or 60 minutes after ACTH administration is diagnostic of PAI [28, 29].

1. In case of hypocortisolism the first step is to evaluate 17-OHP level, because 21-hydroxylase deficiency is the most common cause of PAI in neonates and young infants. Most affected infants have concentrations greater than 35 ng/L and all have concentrations greater than 10 ng/L. If 17-OHP levels are normal, other adrenal insufficiency causes need to be considered. Once 21-hydroxylase deficiency has been ruled out, the most frequent cause of adrenal insufficiency in male neonates are the DAX-1 mutations. Other causes include rare non-virilizing forms of salt-wasting CAH and neonatal adrenal hemorrhage. Adrenal hemorrhage should be suspected in the newborn presenting with adrenal insufficiency and hypovolemic shock in the first week of life and diagnosis is confirmed by abdominal sonography [34]. Screening for possible autoimmune adrenalitis with adrenal autoantibodies and for X-linked adrenoleukodystrophy (ALD) with very long-chain fatty acids (VLCFA) is indicated in infants presenting at an age older than 6 months [10, 42].

2. Normal cortisol, ACTH and 17-OHP levels are consistent with isolated hypoaldosteronism without parallel cortisol deficiency. The next step is to evaluate the renin and aldosterone levels. Elevated PRA and low aldosterone values, particularly an elevated PRA ratio to aldosterone are markers of primary (hyperreninemic) hypoaldosteronism [26]. Low aldosterone and renin concentrations are consistent with hyporeninemic hypoaldosteronism, a diagnosis that is rarely seen in infants. Last, the diagnosis of PHA is established when high aldosterone and renin concentrations are evident in the face of salt wasting and hyperkalemia. In such cases, a renal etiology should be sought as a cause of secondary PHA. Urine culture and renal ultrasonography should be performed in any infant with electrolyte disturbances to exclude infection and obstructive uropathy, even in the absence of fever or other symptoms and signs of pyelonephritis [43].

Antenatal Bartter syndrome should be included in the differential diagnosis of a neonate presenting with hyperkalemia, hyponatremia and hyperreninemic hyperaldosteronism. In general, Bartter syndrome is a group of inherited tubular disorders, characterized renal salt wasting, hypokalemia, metabolic alkalosis and normotensive hyperreninemic hyperaldosteronism. Interestingly, the initial clinical presentation of type II antenatal Bartter syndrome is an important mimic of type 1 PHA. Transient neonatal hyperkalemia, occasionally severe, is observed within the first three weeks of life in the majority of patients, obscuring the initial diagnosis, until the infant becomes hypokalemic. A mean peak plasma potassium level of 9.0 mmol/L (range 6,3–10,5 mmol/L) was documented in 12 neonates with type II antenatal Bartter syndrome and ventricular tachycardia has complicated the clinical course in one of them [44, 45].

5. Treatment

Biochemical confirmation should not delay treatment initiation in acutely sick infants. Infants with a severe salt-wasting present in near-shock to shock and require immediate fluid resuscitation and correction of electrolyte abnormalities. Physicians should keep in mind that a blood sample for the essential hormonal investigations should be collected before any steroid treatment is given to confirm aldosterone insufficiency and to determine the underlying cause. [7, 14, 46]

Hyponatremia is usually long-standing and should be corrected slowly to prevent central pontine myelinolysis [47]. Resolution of hyperkalemia usually occurs rapidly with stress doses of hydrocortisone, due to mineralocorticoid effect. Still, when T-wave elevation is evident on electrocardiogram (ECG), 10% calcium gluconate can be used to stabilize membrane potential. Other specific treatments for hyperkalemia include nebulized salbutamol and intravenous insulin infusion at 1 U of insulin in 5 g dextrose to promote intracellular potassium shifting and kayexalate cation exchange resins to help rid the potassium burden. As a last resort, dialysis can correct hyperkalemia, if T-wave elevation is unrelieved by medical means [10, 16, 47].

Infants with life-threatening salt-wasting crisis are initially treated with parenteral hydrocortisone at stress doses (50–100 mg/m^2 per day divided q 8 h) pending steroid hormone analysis [48]. This approach is not unreasonable given that CAH is a potentially lethal condition if treatment is delayed. Stress doses of hydrocortisone also have adequate mineralocorticoid activity, as 20 mg of intravenous hydrocortisone is equivalent to 100 μg fludrocortisone [10]. Once the infant is stabilized, he or she may be transitioned to oral hydrocortisone and fludrocortisone acetate at doses 30 mg/m^2 per day divided q 8 h and 50 to 100 μg/24 h respectively [49].

Identification of the etiology is crucial to avoid inappropriate prolonged steroid treatment, in case of mineralocorticoid resistance or isolated hypoaldosteronism. Appropriate therapy varies according to the etiology and treatment should be adjusted when the results are available [50].

a. In case of hypocortisolism glucocorticoid replacement therapy is continued according to established guidelines [7]. Maintenance therapy in infants with CAH includes hydrocortisone 12–15 $mg/m^2/d$ and oral fludrocortisone acetate (0.05 to 0.2 mg/24 h). The requirement for sodium in normally growing infants is ~1 mmol/kg per day, the amount provided by human milk. However, in infants with salt-wasting, this amount is insufficient and sodium chloride supplements are recommended at a dose of 1–2 g/d [51].

b. If appropriate cortisol levels are obtained, PAI is excluded and hydrocortisone can be discontinued. Serum aldosterone will further differentiate isolated hypoaldosteronism from PHA. In cases of isolated aldosterone deficiency therapy includes 9a-fludrocortisone and salt supplementation. However, infants with mineralocorticoid resistance will not respond to fludrocortisone treatment. Management of these cases is symptomatic with sodium repletion, ion - exchange resins and treatment of the precipitating cause (e.g. antibiotics for a urinary tract infection) [24].

6. Causes of hypoaldosteronism

Hypoaldosteronism is classified in three large categories, according to their pathophysiology; deficient production by the adrenal glands, aldosterone unresponsiveness and defective stimulation of aldosterone secretion by renin (**Table 1**) [52].

The most common cause of aldosterone deficiency in the first weeks of life, is CAH due to 21-hydroxylase deficiency [38]. However, this diagnosis becomes less likely outside the neonatal period, by which time most cases have been diagnosed, either based on newborn screening or a salt-losing crisis. In boys, once CAH has been ruled out, the most common cause of hypoaldosteronism in early infancy (birth to 2 months) are the DAX-1 mutations, causing AHC. Other defects in aldosterone biosynthesis include ASD, 3β-hydroxysteroid dehydrogenase deficiency, cholesterol side-chain cleavage enzyme deficiency and congenital lipoid hyperplasia due to deficiency of the steroidogenic acute regulatory (StAR) protein. PHA which results from diminished renal tubule responsiveness to aldosterone is another important cause of salt wasting in infancy [6]. Hyporeninemic hypoaldosteronism results in the same metabolic derangements, although this most often presents in adult populations. While rare in infants, the administration of nephrotoxic medications (e.g., ACE inhibitors, nonsteroidal anti-inflammatory drugs) should also be considered [30].

6.1 Deficient aldosterone production by the adrenal glands: hyperreninemic hypoaldosteronism

Where the primary defect is in aldosterone synthesis or release, the serum aldosterone concentration is low with a compensatory increase in PRA. Genetic defects in aldosterone biosynthesis, adrenal destruction and adrenal dysgenesis are the most common reported causes of hyperreninemic hypoaldosteronism [29, 53, 54].

A. Defective production by the adrenal glands: Hyperreninemic hypoaldosteronism (↓ aldosterone - ↑ renin)		
Combined with cortisol insufficiency		Isolated hypoaldosteronism
Genetic disorders	Salt-wasting forms of CAH Adrenal hypoplasia congenita	Aldosterone synthase deficiency
Metabolic disorders	Adrenoleukodystrophy/ Adrenomyeloneuropathy Wolman's disease	
Acquired disorders	Autoimmune adrenalitis Infections Intra-adrenal hemorrhage	Drugs: Heparin, ACE inhibitors, ARBs
B. Aldosterone resistance: Pseudohypoaldosteronism (↑ aldosterone - ↑ renin)		
Primary, due to an inherited receptor defect Secondary (UTI, urinary malformation, drugs)		
C. Defective stimulation by renin: Hyporeninemic hypoaldosteronism (↓ aldosterone - ↓ renin)		
In children with lupus nephritis, post-infectious glomerulonephritis or mild-to-moderate chronic renal insufficiency Drugs: NSAIDs, COX-2 inhibitors, beta-blockers		

Table 1.
Causes of Hypoaldosteronism (CAH: congenital adrenal hyperplasia, ACE: angiotensin-converting enzyme, ARBs: angiotensin II receptor blockers, UTI: urinary tract infection, NSAIDs: nonsteroidal anti-inflammatory drugs, COX-2: cyclooxygenase-2).

6.1.1 Salt-wasting forms of congenital adrenal hyperplasia

CAH is a group of autosomal recessive disorders characterized by cortisol insufficiency due to mutations affecting any of the steroidogenic enzymes required for cortisol synthesis [51]. About 95% of CAH is caused by 21-hydroxylase deficiency, with the aldosterone-deficient form of the disease occurring in approximately 1:20.000 births. Similarly, deficiencies of 3β-hydroxysteroid dehydrogenase type 2 (3βHSD2), steroidogenic acute regulatory protein (StAR) and cholesterol side-chain cleavage enzyme (P450scc) inhibit both cortisol and aldosterone synthesis resulting in adrenal insufficiency with salt loss. Ambiguity of the genitalia in seen in 46,XX with 21OHD and 46,XY with 3βHSD2 deficiency. Paradoxically, 46,XX individuals born with severe 3βHSD2 deficiency can virilize slightly in utero, due to extra-adrenal 3βHSD1 activity. Infants with lipoid CAH (StAR deficiency) have 46,XY sex reversal and normal-appearing female genitalia secondary to a severe defect in Leydig cell steroidogenesis [1, 55]. A detailed review of this topic is beyond the scope of this chapter.

6.1.2 Aldosterone synthase deficiency

ASD is a rare case of hyperreninemic hypoaldosteronism inherited in an autosomal recessive pattern and caused by mutations in the CYP11B2 gene encoding the enzyme aldosterone synthase [10, 56]. The CYP11B2 gene is located on chromosome 8q22p, band q24.3, approximately 40 kb away from the 93% - identical CYP11B1 gene encoding the 11 β-hydroxylase enzyme [57]. Aldosterone synthase catalyzes the three final steps of aldosterone biosynthesis: first the 11-hydroxylation of deoxycorticosterone (DOC) to corticosterone (compound B), then the hydroxylation at position 18 to 18-hydroxycorticosterone (18OHB) and lastly the oxidation at position 18 to aldosterone. According to the relative levels of aldosterone and its precursors, ASD has been subdivided into type 1 and type 2. It is important to note that 11-hydroxylation of DOC is not impaired in either type of ASD because it is also catalyzed by the CYP11B1 isoenzyme, resulting in accumulation of both compound B and DOC [48, 58, 59].

Type 1 ASD, previously known as corticosterone methyloxidase I (CMO I) deficiency is typically characterized by total suppression of aldosterone synthase activity, resulting in impairment of both 18- hydroxylation and 18-oxidation. Thus, patients with ASD 1 have low to normal levels of 18OHB and very low to undetectable levels of aldosterone [58].

Type 2 ASD (CMO II deficiency) results from mutations in CYP11B2 gene that selectively affect the 18-methyl oxidase activity while preserving the 18- hydroxylation of corticosterone, resulting in excessive levels of 18OHB and low to normal levels of aldosterone. Determination of 18OHB-to-aldosterone ratio enables recognition of the site of the enzyme block [10, 19, 60, 61]. Type II ASD is characterized by a markedly (often 100-fold) elevated ratio in either urine or serum [62].

Despite their different biochemical profile, type 1 and type 2 ASD would be better considered a continuous spectrum of the same disease. 18-OHB exhibits minimal biological affinity for the MR and there is considerable overlap between the clinical, hormonal and genotypic features of the two types of the disease [6, 48]. The condition has manifestations ranging from life-threatening salt-wasting crisis in neonates to asymptomatic impairment of statural growth in children. The most common age of onset of major clinical salt wasting is between 1 week and three months of age [6, 58, 60]. Notably, impairment of linear growth may be the sole or the predominant feature in older children [6]. Although fatalities have occasionally occurred, the morbidity of ASD is usually not as severe as that of the salt-wasting forms of CAH, reflecting normal DOC, corticosterone and cortisol synthesis. Moreover, family studies have identified biochemically affected but asymptomatic adults with abnormal ratios of 18-oxygenated steroids [6, 62].

ASD responds well to exogenous mineralocorticoid treatment. Infants will also require NaCl supplements for ongoing electrolyte management. Fludrocortisone doses do not need to be increased with age, since mineralocorticoid sensitivity increases throughout childhood [10, 63]. In the first months of life, fludrocortisone's recommended dosage is 0.05–0.3 mg/day. The dosage might be adjusted to about half of the initial dosage during the second year of life and a third or a quarter during the third year [64]. Mineralocorticoid replacement is typically continued throughout childhood, but is often gradually weaned by adolescence, as patients spontaneously ameliorate their salt-wasting syndrome. Normalization of serum electrolyte concentrations and suppression of PRA towards the normal age-adjusted range seem to represent reasonable objectives in children [6, 11, 63].

6.1.3 Familial hyperreninemic hypoaldosteronism unlinked to the aldosterone synthase (CYP11B2) gene

Isolated hyperreninemic hypoaldosteronism in infancy is usually caused by mutations in CYP11B2 gene. However, there have been several reports of infants with the same clinical picture, in whom no mutations of CYP11B2 were detected. An inherited form of hyperreninemic hypoaldosteronism, distinct from ASD, seems to be the cause and the affected gene(s) remain to be determined [57, 62].

6.1.4 Adrenal hypoplasia congenita

AHC is a rare inherited disorder of adrenal cortex development. It occurs in 2 distinct forms: The X-linked cytomegalic form and the autosomal recessive miniature adult form. In the X-linked or cytomegalic form, the adrenals do not differentiate beyond the fetal stage. They are characterized by an absence of the permanent zone and by abnormally large (cytomegalic) cells. The autosomal recessive or miniature adult form is characterized by small adrenal glands with normal architecture and normal adult zone structure [65–69].

X-linked AHC is caused by a defective NR0B1 (nuclear receptor subfamily 0, group B, member 1) gene [70]. About two thirds of boys with AHC have point mutations and the other one third has gene deletions [71]. The NR0B1 gene encodes the DAX-1 (dosage-sensitive sex reversal, adrenal hypoplasia, critical region on the X chromosome, gene 1) protein on the X-chromosome (Xp21) [69, 72, 73]. DAX-1 is an orphan nuclear receptor expressed in the adrenal cortex, testicular Leydig and Sertoli cells, ovarian theca and granulosa cells, pituitary gonadotropes and hypothalamus [70, 74]. Its actions are mediated by repression of another orphan nuclear receptor, steroidogenic factor 1 (SF-1) and together they regulate the embryological development and subsequent function of these tissues. The prevalence of NR0B1 mutations in the general population has been estimated as 1:70,000–1:600,000 [69, 70, 72–77].

The classic form of X-linked AHC is characterized by three main features: primary adrenal failure, hypogonadotropic hypogonadism (HHG) and infertility [10]. A family history of adrenal failure, unexpected death or HHG, in males in the maternal family is evident in almost 100% of affected individuals [77]. The gold standard for diagnosis of X-linked AHC is genetic testing, showing a deletion or mutation in the NR0B1 (DAX1) gene [69].

6.1.4.1 Primary adrenal failure in X-linked AHC

Classically, a bimodal presentation pattern is seen, with 60% of affected males presenting during the first eight weeks of life and 40% presenting between 1 to 10 years of age with primary adrenal failure or isolated mineralocorticoid deficiency [76, 78]. The initial presentation of X-linked AHC is often a combination of mineral and glucocorticoid deficiency but, especially during the neonatal period, aldosterone deficiency may precede cortisol deficiency at onset [72]. An adult-onset form of X-linked AHC has also been described in 10 males and diagnosis was suspected observing the association of adrenal insufficiency and hypogonadotropic hypogonadism. Variability in the age of onset is evident even among patients of the same family, carrying the same mutation, indicating that epigenetic or environmental factors are also involved in the clinical course [67, 71, 72, 76, 77, 79].

Infants with PAI present with salt-wasting, failure to thrive, hyponatremia, hypoglycemia, and hyperpigmentation. Older individuals may present more insidiously with chronic adrenal insufficiency until a concomitant illness precipitates acute adrenal crisis [71, 73].

A cortisol value within the normal range does not necessarily exclude the diagnosis of AHC and in several cases, children with normal basal cortisol levels presented with clinical adrenal failure shortly after. Glucocorticoid function should be carefully assessed, possibly through a short Synachten test, with sometimes an increase of ACTH level indicating compensated primary adrenal failure [57, 72, 78].

6.1.4.2 Hypogonadotropic hypogonadism in X-linked AHC

X-linked AHC is associated with isolated hypogonadotropic hypogonadism that seems to be the result of both hypothalamic and pituitary dysfunction. The deficit in pituitary hormones is selective for gonadotropins as other hormones' production is normal. HHG usually becomes apparent in adolescence by absent or arrested pubertal development. Progression of puberty beyond Tanner III is extremely uncommon [65, 73, 76, 78, 80, 81].

Cryptorchidism may be present at birth, with at least 10% of infants having unilateral or more frequently bilateral undescended testes [76]. Against expectation, normal minipuberty of infancy with appropriately elevated gonadotropin and testosterone levels has been shown in 2 infants. The maternal uncles sharing the same DAX1 mutation with the infants, were affected by HHG [68, 82].

Other paradoxical features such as macrophallia or transient precocious sexual development have also been described in infancy and childhood, with several mechanisms proposed. Chronic excessive ACTH levels resulting from adrenal insufficiency may stimulate Leydig cells and lead to gonadotropin-independent precocious puberty in some boys with DAX1 gene mutations [70, 83, 84].

6.1.4.3 Complex Glycerol Kinase Deficiency

Males with confirmed X- linked AHC should be evaluated for clinical signs of other diseases mapped in Xp21 because deletions of the NR0B1 gene may also occur along with contiguous gene defects as part of Complex Glycerol Kinase Deficiency (CGKD) [65, 69, 85]. CGKD develops from partial deletion of the Xp21 chromosomal locus involving all or part of the gene for glycerol kinase deficiency (GKD) together with that for AHC and/or Duchenne muscular dystrophy (DMD). Much larger deletions including the ornithine transcarbamylase locus have also been described [86–88]. The syndrome can be both sporadic and familial, and the phenotype varies according to the extension of deleted DNA [89]. Patients with CGKD may show dysmorphic features including prominent eyebrows and forehead and depressed nasal root giving the face an hourglass appearance [86]. Mental impairment is also described, but specific causes have not been clearly defined. The terminal 3′ end of the DMD gene is essential for normal development of the brain and a gene mapped distal to the DMD locus is associated with a form of X-linked mental retardation [89].

6.2 Defective aldosterone action: pseudohypoaldosteronism (PHA)

Syndromes characterized by apparent aldosterone deficiency, despite elevated aldosterone levels are classified as PHA. This may be either primary (PHA type 1 and 2) or secondary (PHA type 3) phenomenon. Primary PHA type 1 is subclassified into two genetically distinct syndromes, that differ in the involvement of aldosterone target organs and the severity of salt wasting: (1) the autosomal dominant (AD) or sporadic form (also called renal form) and (2) the autosomal recessive (AR) or generalized form. The biological characteristics of primary PHA1 and secondary PHA3 are dehydration accompanied by hyponatremia, hyperkalemia, and metabolic acidosis despite high aldosterone levels [27].

In contrast, type 2 PHA (Gordon syndrome or familial hyperkalemic hypertension) is a rare potassium retaining syndrome characterized by hyperkalemia, normal GFR, hypertension, metabolic acidosis, suppressed PRA and variable aldosterone levels. It is caused by mutations affecting WNK1 and WNK4 kinases, as well as Cullin3 (CUL3) and Kelch-like3 (KLHL3) proteins. Comparison between the different types of PHA is presented in **Table 2** [47, 91–93].

6.2.1 Multi-system PHA type 1

Multi-system PHA type 1 (sPHA) is characterized by multiple end-organ resistance to aldosterone and is inherited as an autosomal recessive trait [94]. It is caused by homozygous or compound heterozygous inactivating mutations in the genes encoding the alpha, beta and gamma subunits of the ENaC. Both genes encoding the β- (SNCC1B) and γ-subunits (SNCC1G) are located in 16p12, while the gene encoding the α-subunit (SNCC1A) is located in 12p13 [71, 91, 95–97].

PHA	sPHA type I	rPHA type I	PHA type II	PHA type III
inheritance	AR	AD	AD	not inherited
mutated protein	ENaC	MR	WNK1, WNK4, KLHL3, CUL3	none
patho-physiology	salt loss K^+ retention	salt loss K^+ retention	salt and K^+ retention	salt loss K^+ retention
age of onset	neonatal period	neonatal period early infancy	scholar, adolescence	neonatal period early infancy
blood pressure	hypotension	hypotension	hypertension	hypotension
electrolyte levels	hyponatremia, hyperkalemia	hyponatremia, hyperkalemia	hyperkalemia	hyponatremia, hyperkalemia
PRA	↑	↑	↓	↑
aldosterone	↑	↑	variable	↑
treatment	supplemental sodium, potassium binding resins	supplemental sodium	salt restriction, thiazides	treatment of underlying cause, supplemental sodium
duration	persistent	self-limited	persistent	transient
prognosis	poor	good	good	good

Table 2.
Comparison Between the Different Types of PHA from references [47, 90] (PHA: pseudohypoaldosteronism; AR: autosomal recessive; AD: autosomal dominant; MR: mineralocorticoid receptor; ENaC: epithelial sodium channel; WNK: with-no-lysine (K) kinase; CUL3: Cullin3 and KLHL3: Kelch-like3).

Since the ENaC is expressed in all aldosterone - dependent epithelial tissues (distal part of the nephron, distal colon, salivary ducts, sweat glands, respiratory airway, pulmonary alveoli and nasal mucosa), sPHA is associated with widespread systemic manifestations [91, 98]. The pattern of laboratory abnormalities is diagnostic and shows hyponatremia, hyperkalemia, metabolic acidosis, elevated PRA and aldosterone concentrations. The course of the disease is severe and lifelong treatment is required [92, 99].

In utero, uncontrolled saliuretic fetal polyuria due to mineralocorticoid resistance may lead to polyhydramnios. In the postnatal period, sPHA is characterized by failure to thrive, vomiting and severe dehydration. Affected infants may also have chronic diarrhea, excessive pulmonary secretions, cholelithiasis and recurrent skin rashes [99–101]. Other associated symptoms include chronic discharge of clear liquid from the nose and salt loss from the Meibomian glands of the eyelids [90, 93, 102].

Lower respiratory tract involvement associated with sPHA makes the disease an important mimic of cystic fibrosis. ENaC plays a major role in airway sodium absorption, airway liquid volume and composition [103]. First, the increased volume of intraluminal liquid results in airway narrowing. This is especially evident during infancy and early childhood, when the airway diameter is small. Besides, changes in the airways' ionic composition may compromise normal mucociliary function, predisposing to lower respiratory tract infections. However, children generally do not present after age 5 and do not typically develop *Pseudomonas aeruginosa* lung infections. These features differentiate children with sPHA from those with cystic fibrosis [94, 95, 103–105]. Infants with sPHA sustain recurrent episodes (3–6 per year) of chest congestion, coughing and tachypnea, often associated with fever, wheezing and crackles. It is noteworthy, that respiratory symptoms begin within weeks or months after birth and only two newborns with neonatal respiratory distress syndrome (RDS) and sPHA have been described. Both were premature, one born at 31 weeks of gestation and one born at 36 weeks. Older patients (more than five years of age) have less severe and less frequent respiratory symptoms [47, 90, 98].

Defective ENaC function is also responsible for the high sweat salt concentration of infants with sPHA, making the sweat test an excellent discriminant between the systemic and the renal type of the disease [92]. The high sodium concentration also causes chronic inflammatory changes around and within the sweat ducts resulting in recurrent skin rashes. Cutaneous manifestations of patients with sPHA1 mimic pustular miliaria rubra and are described as discrete erythematous pustules, that worsen during salt-depletion crises and clear spontaneously with stabilization. Interestingly, inflammatory pustules have not been noted in patients with cystic fibrosis. The reason for this is unknown, but may relate to higher sweat salt concentrations in sPHA. Typical sweat chloride concentrations in infants with sPHA range between 110 and 150 mmol/L. In comparison, sweat chloride concentrations higher than 75 mmol/L are reported in patients with cystic fibrosis [99, 102, 106].

Normalization of fluid and electrolyte balance in generalized PHA1 is particularly challenging. Patients are insensitive to mineralocorticoids and require high doses of sodium supplementation (between 20 and 50 mEq/kg/d), together with orally administered ion exchange resins and dietary potassium restriction. Although a slight amelioration is observed with ageing, treatment is mandatory throughout life [102, 107].

6.2.2 Renal PHA type 1

Renal PHA type 1 (rPHA) is an autosomal dominant (AD) disease caused by heterozygous mutations in the NR3C2 gene. The NR3C2 gene located on chromosome 4q31.1 is responsible for encoding the the distal renal tubule's mineralocorticoid receptor [90, 91, 94]. More than 50 different mutations have been identified in this receptor, which lead to renal resistance to aldosterone [71]. The renal type of PHA represents the most frequent form of the disease with a prevalence of 1 per 80.000 newborns [102].

AD-PHA is restricted to the kidneys and clinical symptoms usually remit with age. Although less severe in its course, rPHA has been reported to be associated with high infant mortality rate. In fact, patients with rPHA resemble a striking phenotypic diversity, with a clinical spectrum ranging from asymptomatic to severe PHA. Characteristic of the autosomal dominant form is an affected, symptomatic index case, with family members who are biochemically affected but clinically asymptomatic. Sporadic cases due to de novo mutations have also been reported [90, 97, 108–110].

Patients mainly manifest in early infancy, between 0.5 and 6 months of age, with isolated renal resistance to aldosterone, leading to renal salt loss, hyponatremia, hyperkalemia, metabolic acidosis, failure to thrive and elevated plasma renin and aldosterone concentrations. The main clinical symptom is failure to thrive due to chronic dehydration. Hyperkalemia is generally mild, and metabolic acidosis is not always detectable [93].

In rPHA, 3–20 mEq/kg/daily dose of sodium is sufficient to compensate for the salt loss and is followed by a rapid clinical and biochemical improvement [105]. Potassium-binding resins are rarely needed. Although the primary defect persists for life, improvement usually occurs after the first years of life and sodium supplementation generally becomes unnecessary by 2–3 years of age. Amelioration of the phenotype is attributed to the renal tubule's maturation, autonomous addition of salt to the diet and chronic up-regulation of mineralocorticoid axis. Chronic salt depletion and resultant hyperreninemia possibly stimulates zona glomerulosa leading to the zone's hypertrophy and tertiary hyperaldosteronism. Thus, PRA decreases into normal range, while high plasma aldosterone levels persist into adulthood [90, 102, 105, 108].

6.2.3 Secondary PHA type 3

Secondary PHA in infancy is a transient condition characterized by lack of response to aldosterone in the distal tubule due to obstructive uropathy, VUR and/or UTI [111]. Any kind of urinary tract obstruction, including posterior urethral valves, ureterocele, ureteropelvic junction obstruction and ureterohydronephrosis may lead to PHA [106, 109].

The underlying pathogenesis for secondary aldosterone resistance has not been fully elucidated. Early infancy, however, seems to be the main contributing factor, as the prevalence rate of secondary PHA diminishes considerably after three months of age, with the majority of infants being less than seven months old [106, 110, 112].

Inflammation and production of cytokines is an additional factor contributing to aldosterone resistance. Circulating bacterial endotoxins can directly damage aldosterone receptors, as well as stimulate the intrarenal synthesis of cytokines like prostaglandins, leukotriens, endothelin, interleukin (IL)-1 and thromboxane. Similarily, parenchymal renal damage in case of obstructive uropathy increases the intrarenal expression of tumor necrosis factor-alpha (TNF-α), IL-1, IL-6, transforming growth factor beta-1 (TGF-β1), angiotensin II, endothelin, thromboxane A2 and prostaglandins. These cytokines induce vasoconstriction, reduction of GFR, natriuresis and/or decreased Na^+-K^+-ATPase activity [90, 111].

Secondary PHA is typically an acute condition. Electrolyte imbalance usually resolves after 24–48 hours of intravenous fluid replacement and antibiotic therapy in the case of UTI [113]. However, signs of pseudohypoaldosteronism have been reported to persist even after successful surgery in infants with congenital hydronephrosis, indicative of ongoing distal tubular dysfunction. The required time period for salt supplementation ranges from 3 to 13 months in reported cases, with the youngest infants requiring longer supplementation [25, 106]. If secondary PHA improves with treatment of UTI or obstructive uropathy, further genetic testing for primary PHA1 is not usually suggested [114]. Interestingly, a pathogenic mutation on NR3C2 has been recently identified in an infant with UTI-associated type IV renal tubular acidosis (RTA). Identification of MR or epithelial sodium channel (ENaC) gene polymorphisms in the presence of secondary PHA is suggestive of a possible overlap between primary and secondary type IV RTA [106, 115].

6.3 Defective stimulation by renin: hyporeninemic hypoaldosteronism

Hyporeninemic hypoaldosteronism results from insufficient stimulation of the adrenal gland due to a defect of renin secretion. The syndrome has been especially observed in adults with chronic renal insufficiency due to diabetic nephropathy and rarely in children with lupus nephritis or acute post-infectious glomerulonephritis [15, 116].

Only five infants with hyporeninemic hypoaldosteronism have been reported to date. An 8-month-old boy with chronic kidney disease (CKD) stage 3 caused by tubulointerstitial disease manifested hyperkalemia (potassium = 7.1 mEq/L) with normal GFR in the context of hyporeninemic hypoaldosteronism [117]. Hyporeninemic hypoaldosteronism has also been reported in a 3-month-old boy with severe psychomotor retardation and growth failure and a 5-month-old boy with severe mental retardation lactic acidosis and deafness [116]. Finally, the report of two male siblings, presenting with hyporeninemic hypoaldosteronism at the age of 12 and 2 months suggested a congenital primary defect [118].

7. Conclusions

Although rare, hypoaldosteronism is a potential cause of neonatal morbidity and mortality due to electrolyte disturbances and hypovolemia. Early diagnosis and treatment represent a major challenge for pediatricians, who should be aware of this condition either as isolated hypoaldosteronism or in the context of PAI. A deeper understanding of the etiology of hypoaldosteronism is crucial, to improve care of affected infants [1, 119].

Author details

Elpis-Athina Vlachopapadopoulou* and Myrto Bonataki
Department of Endocrinology, Growth and Development, 'P. & A. Kyriakou' Children's Hospital, Athens, Greece

*Address all correspondence to: elpis.vl@gmail.com

References

[1] White PC. Abnormalities of aldosterone synthesis and action in children. Curr Opin Pediatr. 1997 Aug;9(4):424-30. doi: 10.1097/00008480-199708000-00019. PMID: 9300202.

[2] Solomon N, Many A, Orbach R, Mandel D, Shinar S. Maternal and neonatal hyponatremia during labor: a case series. J Matern Neonatal Med 2019;32:2711-5. https://doi.org/10.1080/14767058.2018.1446517.

[3] Laetitia M, Eric P, Laurence FLH, Francois P, Claudine C, Pascal B, et al. Physiological partial aldosterone resistance in human newborns. Pediatr Res 2009;66:323-8. https://doi.org/10.1203/PDR.0b013e3181b1bbec.

[4] Rodríguez-Soriano, J., Ubetagoyena, M. & Vallo, A. Transtubular potassium concentration gradient: a useful test to estimate renal aldosterone bio-activity in infants and children. Pediatr Nephrol 4, 105-110 (1990). https://doi.org/10.1007/BF00858819

[5] Martinerie L, Pussard E, Yousef N, Cosson C, Lema I, Husseini K, et al. Aldosterone-signaling defect exacerbates sodium wasting in very preterm neonates: The premaldo study. J Clin Endocrinol Metab 2015;100:4074-81. https://doi.org/10.1210/jc.2015-2272.

[6] Veldhuis JD, Melby JC. Isolated aldosterone deficiency in man: acquired and inborn errors in the biosynthesis or action of aldosterone. Endocr Rev. 1981 Fall;2(4):495-517. doi: 10.1210/edrv-2-4-495. PMID: 6273148.

[7] Bizzarri C, Pedicelli S, Cappa M, Cianfarani S. Water balance and "salt wasting" in the first year of life: The role of aldosterone-signaling defects. Horm Res Paediatr 2016;86:143-53. https://doi.org/10.1159/000449057.

[8] Storey C, Dauger S, Deschenes G, Heneau A, Baud O, Carel JC, et al. Hyponatremia in children under 100 days old: incidence and etiologies. Eur J Pediatr 2019;178:1353-61. https://doi.org/10.1007/s00431-019-03406-8.

[9] Delgado MM, Rohatgi R, Khan S, Holzman IR, Satlin LM. Sodium and potassium clearances by the maturing kidney: Clinical-molecular correlates. Pediatr Nephrol 2003;18:759-67. https://doi.org/10.1007/s00467-003-1178-1.

[10] Claire R. Hughes EM and JCA. The Adrenal Cortex and Its Disorders. Brook's Clin. Pediatr. Endocrinol. 7th ed., Chichester: Wiley; 2020, p. 335-407. https://doi.org/10.1088/1751-8113/44/8/085201.

[11] Rösler A. The natural history of salt-wasting disorders of adrenal and renal origin. J Clin Endocrinol Metab 1984;59:689-700. https://doi.org/10.1210/jcem-59-4-689.

[12] Moritz ML, Ayus JC. Disorders of water metabolism in children: hyponatremia and hypernatremia. Pediatr Rev 2002;23:371-80. https://doi.org/10.1542/pir.23-11-371.

[13] Moritz ML, Ayus JC. New aspects in the pathogenesis, prevention, and treatment of hyponatremic encephalopathy in children. Pediatr Nephrol 2010;25:1225-38. https://doi.org/10.1007/s00467-009-1323-6.

[14] Powers KS. Dehydration: Isonatremic, hyponatremic, and hypernatremic recognition and management. Pediatr Rev 2015;36: 274-85. https://doi.org/10.1542/pir.36-7-274.

[15] Rodríguez-Soriano, J. Potassium homeostasis and its disturbances in children. Pediatr Nephrol 9, 364-374

(1995). https://doi.org/10.1007/BF02254217

[16] Lehnhardt A, Kemper MJ. Pathogenesis, diagnosis and management of hyperkalemia. Pediatr Nephrol 2011;26:377-84. https://doi.org/10.1007/s00467-010-1699-3.

[17] Haycock, G.B. The influence of sodium on growth in infancy. Pediatr Nephrol 7, 871-875 (1993). https://doi.org/10.1007/BF01213376.

[18] Stritzke A, Thomas S, Amin H, Fusch C, Lodha A. Renal consequences of preterm birth. Mol Cell Pediatr 2017;4. https://doi.org/10.1186/s40348-016-0068-0.

[19] Giapros VI, Tsatsoulis AA, Drougia EA, Kollios KD, Siomou EC, Andronikou SK d. Rare causes of acute hyperkalemia in the 1st week of life. Three case reports. Pediatr Nephrol 2004;19:1046-9. https://doi.org/10.1007/s00467-004-1500-6.

[20] Bassam A. Atiyeh SSD and ABG. Evaluation of Renal Function During Childhood. Pediatr Rev 1996;17:175-80. https://doi.org/10.1542/pir.17-5-175.

[21] Sulyok E, Nemeth M, Tenyi I, Csaba IF, Varga F, Györy E, et al. Relationship between maturity, electrolyte balance and the function of the renin-angiotensin-aldosterone system in newborn infants. Biol Neonate 1979;35:60-5. https://doi.org/10.1159/000241154.

[22] Krishnappa V, Ross JH, Kenagy DN, Raina R. Secondary or Transient Pseudohypoaldosteronism Associated With Urinary Tract Anomaly and Urinary Infection: A Case Report. Urol Case Reports 2016;8:61-2. https://doi.org/10.1016/j.eucr.2016.07.001.

[23] Bizzarri C, Olivini N, Pedicelli S, Marini R, Giannone G, Cambiaso P, et al. Congenital primary adrenal insufficiency and selective aldosterone defects presenting as salt-wasting in infancy: A single center 10-year experience. Ital J Pediatr 2016;42:1-8. https://doi.org/10.1186/s13052-016-0282-3.

[24] Manikam L, Cornes MP, Kalra D, Ford C, Gama R. Transient pseudohypoaldosteronism masquerading as congenital adrenal hyperplasia. Ann Clin Biochem 2011;48:380-2. https://doi.org/10.1258/acb.2011.010264.

[25] Bertini A, Milani GP, Simonetti GD, Fossali EF, Faré PB, Bianchetti MG, et al. Na+, K+, Cl−, acid–base or H2O homeostasis in children with urinary tract infections: a narrative review. Pediatr Nephrol 2016;31:1403-9. https://doi.org/10.1007/s00467-015-3273-5.

[26] Antal Z, Zhou P. Congenital adrenal hyperplasia: diagnosis, evaluation, and management. Pediatr Rev. 2009 Jul;30(7):e49-57. doi: 10.1542/pir.30-7-e49. PMID: 19570920.

[27] Mastrandrea LD, Martin DJ, Springate JE. Clinical and biochemical similarities between reflux/obstructive uropathy and salt-wasting congenital adrenal hyperplasia. Clin Pediatr (Phila) 2005;44:809-12. https://doi.org/10.1177/000992280504400911.

[28] Shulman DI, Palmert MR, Kemp SF. Adrenal insufficiency: Still a cause of morbidity and death in childhood. Pediatrics 2007;119. https://doi.org/10.1542/peds.2006-1612.

[29] Auron M, Raissouni N. Adrenal Insufficiency. Pediatr Rev 2015;36:92-103.

[30] Nandagopal R, Vaidyanathan P, Kaplowitz P. Transient Pseudohypoaldosteronism due to Urinary Tract Infection in Infancy: A Report of 4 Cases. Int J Pediatr Endocrinol 2009;2009:1-4. https://doi.org/10.1155/2009/195728.

[31] Abraham MB, Larkins N, Choong CS, Shetty VB. Transient pseudohypoaldosteronism in infancy secondary to urinary tract infection. J Paediatr Child Health 2017;53:458-63. https://doi.org/10.1111/jpc.13481.

[32] Nalcacıoglu H, Genc G, Meydan BC, Ozkaya O. Hyperkalaemia in a female patient with systemic lupus erythematosus: answers. Pediatr Nephrol 2012;27:1501-3. https://doi.org/10.1007/s00467-011-2043-2.

[33] Al-Alwan I, Navarro O, Daneman D, Daneman A. Clinical utility of adrenal ultrasonography in the diagnosis of congenital adrenal hyperplasia. J Pediatr. 1999 Jul;135(1):71-5. doi: 10.1016/s0022-3476(99)70330-6. PMID: 10393607.

[34] Toti MS, Ghirri P, Bartoli A, Caputo C, Laudani E, Masoni F, et al. Adrenal hemorrhage in newborn: How, when and why- from case report to literature review. Ital J Pediatr 2019;45:1-8. https://doi.org/10.1186/s13052-019-0651-9.

[35] Tasian GE, Copp HL, Baskin LS. Diagnostic imaging in cryptorchidism: Utility, indications, and effectiveness. J Pediatr Surg 2011;46:2406-13. https://doi.org/10.1016/j.jpedsurg.2011.08.008.

[36] Miller WL. Disorders in the initial steps of steroid hormone synthesis. J Steroid Biochem Mol Biol 2017;165:18-37. https://doi.org/10.1016/j.jsbmb.2016.03.009.

[37] Atmis B, Turan İ, Melek E, Bayazit AK. An infant with hyponatremia, hyperkalemia, and metabolic acidosis associated with urinary tract infection: Questions. Pediatr Nephrol 2019;34:1739-41. https://doi.org/10.1007/s00467-019-04254-2.

[38] Antal Z, Zhou P. Addison Disease. Pediatr Rev 2009;30:491-3. https://doi.org/10.7748/phc.16.8.4.s2.

[39] Memoli E, Lava SAG, Bianchetti MG, Vianello F, Agostoni C, Milani GP. Prevalence, diagnosis, and management of secondary pseudohypoaldosteronism. Pediatr Nephrol 2020;35:713-4. https://doi.org/10.1007/s00467-019-04419-z.

[40] Filipa Almeida, Susana Lopes, Margarida Figueiredo, Filipe Oliveira SS, Sequeira and A. Dehydration and Electrolyte Abnormalities in an 11-year-old Boy. Pediatr Rev 2017;38:441-4. https://doi.org/10.1542/pir.2015-0165.

[41] Regelmann MO, Kamboj MK, Miller BS, Nakamoto JM, Sarafoglou K, Shah S, et al. Adrenoleukodystrophy: Guidance for adrenal surveillance in males identified by newborn screen. J Clin Endocrinol Metab 2018;103:4324-31. https://doi.org/10.1210/jc.2018-00920.

[42] Huffnagel IC, Laheji FK, Aziz-Bose R, Tritos NA, Marino R, Linthorst GE, et al. The natural history of adrenal insufficiency in X-linked adrenoleukodystrophy: An international collaboration. J Clin Endocrinol Metab 2019;104:118-26. https://doi.org/10.1210/jc.2018-01307.

[43] Pai B, Shaw N, Högler W. Salt-losing crisis in infants-not always of adrenal origin. Eur J Pediatr 2012;171:317-21. https://doi.org/10.1007/s00431-011-1541-3.

[44] Brochard K, Boyer O, Blanchard A, Loirat C, Niaudet P, MacHer MA, et al. Phenotype-genotype correlation in antenatal and neonatal variants of Bartter syndrome. Nephrol Dial Transplant 2009;24:1455-64. https://doi.org/10.1093/ndt/gfn689.

[45] Finer G, Shalev H, Birk OS, Galron D, Jeck N, Sinai-Treiman L, et al. Transient neonatal hyperkalemia in the antenatal (ROMK defective) Bartter syndrome. J Pediatr 2003;142:318-23. https://doi.org/10.1067/mpd.2003.100.

[46] Bornstein SR, Allolio B, Arlt W, Barthel A, Don-Wauchope A, Hammer GD, et al. Diagnosis and Treatment of Primary Adrenal Insufficiency: An Endocrine Society Clinical Practice Guideline. J Clin Endocrinol Metab 2016;101:364-89. https://doi.org/10.1210/jc.2015-1710.

[47] Sopfe J, Simmons JH. Failure to thrive, hyponatremia, and hyperkalemia in a neonate. Pediatr Ann 2013;42:74-9. https://doi.org/10.3928/00904481-20130426-09.

[48] Hui E, Yeung MCW, Cheung PT, Kwan E, Low L, Tan KCB, et al. The clinical significance of aldosterone synthase deficiency: Report of a novel mutation in the CYP11B2 gene. BMC Endocr Disord 2014;14:1-5. https://doi.org/10.1186/1472-6823-14-29.

[49] Spitzer MD. Index of suspicion. Pediatr Rev. 2001 Dec;22(12):413-8. doi: 10.1542/pir.22-12-413. PMID: 11731682.

[50] Young WF. Etiology, diagnosis, and treatment of hypoaldosteronism (type 4 RTA). UpToDate 2020:1-17.

[51] Speiser PW, Arlt W, Auchus RJ, Baskin LS, Conway GS, Merke DP, et al. Congenital adrenal hyperplasia due to steroid 21-hydroxylase deficiency: An endocrine society* clinical practice guideline. vol. 103. 2018. https://doi.org/10.1210/jc.2018-01865.

[52] Arai K, Papadopoulou-Marketou N, Chrousos GP. Aldosterone Deficiency and Resistance. [Updated 2020 Nov 24]. In: Feingold KR, Anawalt B, Boyce A, et al., editors. Endotext [Internet]. South Dartmouth (MA): MDText.com, Inc.; 2000-. Available from: https://www.ncbi.nlm.nih.gov/books/NBK279079/

[53] Rajbhandari P, Mandelia C, Janjua HS, Selvakumar PKC, Krishna S. A 9-day-old with weight loss and diarrhea. Pediatrics 2017;139. https://doi.org/10.1542/peds.2016-2953.

[54] Rajkumar V, Waseem M. Hypoaldosteronism. 2021 Feb 10. In: StatPearls [Internet]. Treasure Island (FL): StatPearls Publishing; 2021 Jan–. PMID: 32310452.

[55] Miller WL. Steroidogenesis: Unanswered Questions. Trends Endocrinol Metab 2017;28:771-93. https://doi.org/10.1016/j.tem.2017.09.002.

[56] Lages A de S, Vale B, Oliveira P, Cardoso R, Dinis I, Carrilho F, et al. Congenital hyperreninemic hypoaldosteronism due to aldosterone synthase deficiency type i in a portuguese patient – Case report and review of literature. Arch Endocrinol Metab 2019;63:84-8. https://doi.org/10.20945/2359-3997000000107.

[57] Kayes-Wandover KM, Tannin GM, Shulman D, Peled D, Jones KL, Karaviti L, White PC. Congenital hyperreninemic hypoaldosteronism unlinked to the aldosterone synthase (CYP11B2) gene. J Clin Endocrinol Metab. 2001 Nov;86(11):5379-82. doi: 10.1210/jcem.86.11.8005. PMID: 11701710.

[58] Kayes-Wandover KM, Schindler REL, Taylor HC, White PC. Type 1 Aldosterone Synthase Deficiency Presenting in a Middle-Aged Man 1 . J Clin Endocrinol Metab 2001;86:1008-12. https://doi.org/10.1210/jcem.86.3.7326.

[59] Nieto-Vega FA, Martín-Masot R, Rodríguez-Azor B, Martínez-Rivera V, Herrador-López M, Navas-López VM. A rare cause of chronic hyponatremia in an infant: Answers. Pediatr Nephrol 2021;36:79-81. https://doi.org/10.1007/s00467-020-04615-2.

[60] Jessen CL, Christensen JH, Birkebæk NH, Rittig S. Homozygosity for a mutation in the CYP11B2 gene in an infant with congenital corticosterone methyl oxidase deficiency type II. Acta Paediatr Int J Paediatr 2012;101. https://

doi.org/10.1111/j.1651-2227.2012.02823.x.

[61] Mutlu GY, Bilge, Mehmet Taşdemir, Nuray Uslu Kızılkan, Tülay Güran, Şükrü Hatun HKI. A rare cause of chronic hyponatremia in an infant: Answers. Pediatr Nephrol 2019. https://doi.org/10.1007/s00467-020-04608-1.

[62] White PC. Aldosterone synthase deficiency and related disorders. Mol Cell Endocrinol 2004;217:81-7. https://doi.org/10.1016/j.mce.2003.10.013.

[63] Visclosky T, Kendall A, Biary N. Failure to Thrive in a 2-month-old Boy. Pediatr Rev 2020;41:642-4. https://doi.org/10.1542/pir.2019-0013.

[64] Stapenhorst L. 9A-Fluorohydrocortisone Therapy in Aldosterone Synthase Deficiency. Pediatr Nephrol 2005;20:839. https://doi.org/10.1007/s00467-005-1824-x.

[65] Kletter GB, Gorski JL, Kelch RP. Congenital adrenal hypoplasia and isolated gonadotropin deficiency. Trends Endocrinol Metab 1991;2:123-8. https://doi.org/10.1016/1043-2760(91)90001-4.

[66] Vilain E, Merrer MLE, Lecointre C, Desangles F, Kay MA, Maroteaux P, et al. IMAGe, a new clinical association of Intrauterine growth retardation, Metaphyseal dysplasia, adrenal hypoplasia congenita, and Genital anomalies. J Clin Endocrinol Metab 1999;84:4335-40. https://doi.org/10.1210/jcem.84.12.6186.

[67] Muscatelli F, Strom TM, Walker AP, Zanaria E, Récan D, Meindl A, et al. Mutations in the DAX-1 gene give rise to both X-linked adrenal hypoplasia congenita and hypogonadotropic hypogonadism. Nature 1994;372:672-6. https://doi.org/10.1038/372672a0.

[68] Kaiserman KB, Nakamoto JM, Geffner ME, McCabe ERB. Minipuberty of infancy and adolescent pubertal function in adrenal hypoplasia congenita. J Pediatr 1998;133:300-2. https://doi.org/10.1016/S0022-3476(98)70242-2.

[69] Dhamne C, Adesida R, Agrawal V. Case 1: A Term Male Infant with Hypoglycemia and Hyperbilirubinemia as Initial Presentation. Neoreviews 2007;8:e301-3. https://doi.org/10.1542/neo.8-7-e301.

[70] Landau Z, Hanukoglu A, Sack J, Goldstein N, Weintrob N, Eliakim A, et al. Clinical and genetic heterogeneity of congenital adrenal hypoplasia due to NR0B1 gene mutations. Clin Endocrinol (Oxf) 2010;72:448-54. https://doi.org/10.1111/j.1365-2265.2009.03652.x.

[71] Sperling MA. Adrenal cortex and its disorders. Pediatr. Endocrinol., n.d., p. 471-526.

[72] Iughetti L, Lucaccioni L, Bruzzi P, Ciancia S, Bigi E, Madeo SF, et al. Isolated hypoaldosteronism as first sign of X-linked adrenal hypoplasia congenita caused by a novel mutation in NR0B1/DAX-1 gene: A case report. BMC Med Genet 2019;20:1-8. https://doi.org/10.1186/s12881-019-0834-7.

[73] Jain S, Karaviti LP. A 4-year-old Boy with Recurrent Vomiting. Pediatr Rev 2019;40:425-7. https://doi.org/10.1542/pir.2017-0280.

[74] Peter M, Viemann M, Partsch CJ, Sippell WG. Congenital adrenal hypoplasia: Clinical spectrum, experience with hormonal diagnosis, and report on new point mutations of the DAX-1 gene. J Clin Endocrinol Metab 1998;83:2666-74. https://doi.org/10.1210/jcem.83.8.5027.

[75] Tabarin A, Achermann JC, Recan D, Bex V, Bertagna X, Christin-Maitre S, et al. A novel mutation in DAX1 causes delayed-onset adrenal insufficiency and incomplete hypogonadotropic

hypogonadism. J Clin Invest 2000;105:321-8. https://doi.org/10.1172/JCI7212.

[76] Kyriakakis N, Shonibare T, Kyaw-Tun J, Lynch J, Lagos CF, Achermann JC, et al. Late-onset X-linked adrenal hypoplasia (DAX-1, NR0B1): two new adult-onset cases from a single center. Pituitary 2017;20:585-93. https://doi.org/10.1007/s11102-017-0822-x.

[77] Lin L, Gu WX, Ozisik G, To WS, Owen CJ, Jameson JL, et al. Analysis of DAX1 (NR0B1) and steroidogenic factor-1 (NR5A1) in children and adults with primary adrenal failure: Ten years' experience. J Clin Endocrinol Metab 2006;91:3048-54. https://doi.org/10.1210/jc.2006-0603.

[78] Achermann JC, Meeks JJ, Jameson JL. X-linked adrenal hypoplasia congenita and DAX-1. Endocrinologist 2000;10:289-99. https://doi.org/10.1097/00019616-200010050-00003.

[79] Yanase T, Takayanagi R, Oba K, Nishi Y, Ohe K, Nawata H. New mutations of DAX-1 genes in two Japanese patients with X-linked congenital adrenal hypoplasia and hypogonadotropic hypogonadism. J Clin Endocrinol Metab. 1996 Feb;81(2):530-5. doi: 10.1210/jcem.81.2.8636263. PMID: 8636263.

[80] Habiby RL, Boepple P, Nachtigall L, Sluss PM, Crowley WF, Jameson JL. Adrenal hypoplasia congenita with hypogonadotropic hypogonadism: Evidence that DAX-1 mutations lead to combined hypothalamic and pituitary defects in gonadotropin production. J Clin Invest 1996;98:1055-62. https://doi.org/10.1172/JCI118866.

[81] Wiltshire E, Couper J, Rodda C, Jameson L, Achermann JC. Variable presentation of X-linked adrenal hypoplasia congenita. J Pediatr Endocrinol Metab 2001;14:1093-6. https://doi.org/10.1515/jpem-2001-0804.

[82] Takahashi T, Shoji Y, Shoji Y, Haraguchi N, Takahashi I, Takada G. Active hypothalamic-pituitary-gonadal axis in an infant with X-linked adrenal hypoplasia congenita. J Pediatr 1997;130:485-8. https://doi.org/10.1016/S0022-3476(97)70217-8.

[83] Domenice S, Latronico AC, Brito VN, Arnhold IJP, Kok F, Mendonca BB. Clinical case seminar: Adrenocorticotropin-dependent precocious puberty of testicular origin in a boy with X-linked adrenal hypoplasia congenita due to a novel mutation in the DAX1 gene. J Clin Endocrinol Metab 2001;86:4068-71. https://doi.org/10.1210/jcem.86.9.7816.

[84] Guzzetti C, Bizzarri C, Pisaneschi E, Mucciolo M, Bellacchio E, Ibba A, et al. Next-Generation Sequencing Identifies Different Genetic Defects in 2 Patients with Primary Adrenal Insufficiency and Gonadotropin-Independent Precocious Puberty. Horm Res Paediatr 2018;90:203-11. https://doi.org/10.1159/000492496.

[85] Ramanjam V, Delport S, Wilmshurst JM. The diagnostic difficulties of complex glycerol kinase deficiency. J Child Neurol 2010;25:1269-71. https://doi.org/10.1177/0883073809357240.

[86] Korkut S, Baştuğ O, Raygada M, Hatipoğlu N, Kurtoğlu S, Kendirci M, et al. Complex glycerol kinase deficiency and adrenocortical insufficiency in two neonates. JCRPE J Clin Res Pediatr Endocrinol 2016;8:468-71. https://doi.org/10.4274/jcrpe.2539.

[87] Scheuerle A, Greenberg F, McCabe ERB. Dysmorphic features in patients with complex glycerol kinase deficiency. J Pediatr 1995;126:764-7.

https://doi.org/10.1016/S0022-3476(95)70409-4.

[88] Mazzawi M, Reed D, Agrawal S, Rauch DA. CAH, or is it? differentiating the causes of hypoaldosteronism in pediatrics. Pediatrics 2019;144.

[89] Zaffanello M, Zamboni G, Tonin P, Solero GP, Tatò L. Complex glycerol kinase deficiency leads to psychomotor and body-growth failure. J Paediatr Child Health 2004;40:237-40. https://doi.org/10.1111/j.1440-1754.2004.00346.x.

[90] Kostakis ID, Cholidou KG, Perrea D. Syndromes of impaired ion handling in the distal nephron: Pseudohypoaldosteronism and familial hyperkalemic hypertension. Hormones 2012;11:31-53. https://doi.org/10.1007/bf03401536.

[91] Mittal A, Khera D, Vyas V, Choudhary B, Singh K. Dangerous hyperkalemia in a newborn: Answers. Pediatr Nephrol. 2019 May;34(5):813-815. doi: 10.1007/s00467-018-4102-4. Epub 2018 Oct 31. PMID: 30382335.

[92] Bonny O, Knoers N, Monnens L, Rossier BC. A novel mutation of the epithelial Na + channel causes type 1 pseudohypoaldosteronism. Pediatr Nephrol 2002:804-8. https://doi.org/10.1007/s00467-002-0945-8.

[93] Riepe FG. Pseudohypoaldosteronism. Horm. Resist. Hypersensitivity From Genet. to Clin. Manag., vol. 24, 2013, p. 86-95. https://doi.org/10.1159/000342508.

[94] Hanukoglu A, Bistritzer T, Rakover Y, Mandelberg A. Pseudohypoaldosteronism with increased sweat and saliva electrolyte values and frequent lower respiratory tract infections mimicking cystic fibrosis. J Pediatr 1994;125:752-5. https://doi.org/10.1016/S0022-3476(06)80176-9.

[95] Rodríguez-Soriano J. New insights into the pathogenesis of renal tubular acidosis--from functional to molecular studies. Pediatr Nephrol. 2000 Oct;14(12):1121-36. doi: 10.1007/s004670000407. PMID: 11045400.

[96] Nyström AM, Bondeson ML, Skanke N, Mårtensson J, Strömberg B, Gustafsson J, et al. A Novel Nonsense Mutation of the Mineralocorticoid Receptor Gene in a Swedish Family with Pseudohypoaldosteronism Type I (PHA1). J Clin Endocrinol Metab 2004;89:227-31. https://doi.org/10.1210/jc.2003-030762.

[97] Geller DS, Zhang J, Zennaro MC, Vallo-Boado A, Rodriguez-Soriano J, Furu L, et al. Autosomal dominant pseudohypoaldosteronism type 1: Mechanisms, evidence for neonatal lethality, and phenotypic expression in adults. J Am Soc Nephrol 2006;17:1429-36. https://doi.org/10.1681/ASN.2005111188.

[98] Malagon-Rogers M. A patient with pseudohypoaldosteronism type 1 and respiratory distress syndrome. Pediatr Nephrol. 1999 Aug;13(6):484-6. doi: 10.1007/s004670050643. PMID: 10452275.

[99] Urbatsch A, Paller AS. Pustular miliaria rubra: A specific cutaneous finding of type I pseudohypoaldosteronism. Pediatr Dermatol 2002;19:317-9. https://doi.org/10.1046/j.1525-1470.2002.00090.x.

[100] Liu CC, Lin SH, Sung CC, Lin CM. A neonate with poor weight gain and hyperkalemia: Answers. Pediatr Nephrol 2017;32:73-5. https://doi.org/10.1007/s00467-015-3281-5.

[101] Greenberg D, Abramson O, Phillip M. Fetal pseudohypoaldosteronism: another cause of hydramnios. Acta Pædiatrica 1995;84:582-4. https://doi.org/10.1111/j.1651-2227.1995.tb13701.x.

[102] Zennaro MC, Hubert EL, Fernandes-Rosa FL. Aldosterone resistance: Structural and functional considerations and new perspectives. Mol Cell Endocrinol 2012;350:206-15. https://doi.org/10.1016/j.mce.2011.04.023.

[103] Kerem E, Bistritzer T, Hanukoglu A, Hofmann T, Zhou Z, Bennett W, MacLaughlin E, Barker P, Nash M, Quittell L, Boucher R, Knowles MR. Pulmonary epithelial sodium-channel dysfunction and excess airway liquid in pseudohypoaldosteronism. N Engl J Med. 1999 Jul 15;341(3):156-62. doi: 10.1056/NEJM199907153410304. PMID: 10403853.

[104] Magill SB. Pathophysiology, diagnosis, and treatment of mineralocorticoid disorders. Compr Physiol 2014;4:1083-119. https://doi.org/10.1002/cphy.c130042.

[105] Conversano E, Romano S, Taddio A, Faletra F, Zanon D, Barbi E, et al. When salt is needed to grow: Answers. Pediatr Nephrol 2020:1-2. https://doi.org/10.1007/s00467-020-04647-8.

[106] Belot A, Ranchin B, Fichtner C, Pujo L, Rossier BC, Liutkus A, et al. Pseudohypoaldosteronisms, report on a 10-patient series. Nephrol Dial Transplant 2008;23:1636-41. https://doi.org/10.1093/ndt/gfm862.

[107] Bistritzer T, Lahat E, Eshel G, Barr J, Hanukoglu A, Aladjem M. Severe pseudohypoaldosteronism in a pair of twins not associated with hydramnios. Pediatr Nephrol. 1996 Aug;10(4):438-41. doi: 10.1007/s004670050134. PMID: 8865238.

[108] Hanukoglu A. Type I pseudohypoaldosteronism includes two clinically and genetically distinct entities with either renal or multiple target organ defects. J Clin Endocrinol Metab. 1991 Nov;73(5):936-44. doi: 10.1210/jcem-73-5-936. PMID: 1939532.

[109] Kuhnle U. Pseudohypoaldosteronism: Mutation found, problem solved? Mol Cell Endocrinol 1997;133:77-80. https://doi.org/10.1016/S0303-7207(97)00149-4.

[110] Bowden SA, Cozzi C, Hickey SE, Thrush DL, Astbury C, Nuthakki S. Autosomal Dominant Pseudohypoaldosteronism Type 1 in an Infant with Salt Wasting Crisis Associated with Urinary Tract Infection and Obstructive Uropathy. Case Rep Endocrinol 2013;2013:1-5. https://doi.org/10.1155/2013/524647.

[111] Kibe T, Sobajima T, Yoshimura A, Uno Y, Wada N, Ueta I. Secondary pseudohypoaldosteronism causing cardiopulmonary arrest and cholelithiasis. Pediatr Int 2014;56:270-2. https://doi.org/10.1111/ped.12267.

[112] Gil-Ruiz MA, Alcaraz AJ, Marañón RJ, Navarro N, Huidobro B, Luque A. Electrolyte disturbances in acute pyelonephritis. Pediatr Nephrol 2012;27:429-33. https://doi.org/10.1007/s00467-011-2020-9.

[113] Delforge X, Kongolo G, Cauliez A, Braun K, Haraux E, Buisson P. Transient pseudohypoaldosteronism: a potentially severe condition affecting infants with urinary tract malformation. J Pediatr Urol 2019;15:265.e1-265.e7. https://doi.org/10.1016/j.jpurol.2019.03.002.

[114] Atmis B, Turan İ, Melek E, Bayazit AK. An infant with hyponatremia, hyperkalemia, and metabolic acidosis associated with urinary tract infection: Answers. Pediatr Nephrol 2019;34:1739-41. https://doi.org/10.1007/s00467-019-04254-2.

[115] Tseng MH, Huang JL, Huang SM, Tsai JD, Wu TW, Fan WL, et al. Clinical features, genetic background, and outcome in infants with urinary tract

infection and type IV renal tubular acidosis. Pediatr Res 2020;87:1251-5. https://doi.org/10.1038/s41390-019-0727-7.

[116] Rodríguez-Soriano J, Vallo A. Renal tubular hyperkalaemia in childhood. Pediatr Nephrol 1988;2:498-509. https://doi.org/10.1007/BF00853448.

[117] Lee BH, Kang HG, Choi Y. Hyporeninemic hypoaldosteronism in a child with chronic kidney disease - Is this condition renoprotective? Pediatr Nephrol 2009;24:1771-2. https://doi.org/10.1007/s00467-009-1121-1.

[118] Landier F, Guyene TT, Boutignon H, Nahoul K, Corvol P, Jean-Claude J. Hyporeninemic Hypoaldosteronism in Infancy: A Familial Disease. J Clin Endocrinol Metab 1984;58:143-8. https://doi.org/10.1210/jcem-58-1-143.

[119] Papailiou S, Vlachopapadopoulou EA, Sertedaki A, Maritsi D, Syggelos N, Syggelou A. Aldosterone synthase deficiency type II: An unusual presentation of the first Greek case reported with confirmed genetic analysis. Endocr Regul 2020;54:227-9. https://doi.org/10.2478/enr-2020-0025.

Chapter 8

The Role of Renin Angiotensin Aldosterone System in the Pathogenesis and Pathophysiology of COVID-19

Ozlem G. Sahin

Abstract

The novel coronavirus also known as severe acute respiratory syndrome coronavirus 2 (SARS-CoV-2) whose origin is still having uncertainties related to the existence of an intermediate host, has created the currently ongoing pandemic of coronavirus disease 2019. (COVID-19) The binding assays of SARS-CoV-2 spike protein receptor binding domain disclosed enhanced affinity with human angiotensin II-converting enzyme receptor (hACE2) comparing to the bat ACE2 receptors. ACE2, is an essential component of the regulatory mechanism of the renin-angiotensin-aldosterone system, (RAAS) and this pathway is considered to interact with the pathophysiology of COVID-19. In this chapter, we will discuss the key role of RAAS in the pathogenesis of SARS-CoV-2.

Keywords: ACE2, RAAS, SARS-CoV-2, COVID-19, Ang II, ADAM17

1. Introduction

1.1 The pathogenic interaction of SARS-CoV-2 and renin: angiotensin: aldosterone system

Coronaviruses (CoVs) belong to the family of Coronaviridae which is further divided into four genera as Alphacoronavirus, (α-CoV) Betacoronavirus, (β-CoV) Gammacoronavirus, (γ-CoV) and Deltacoronavirus. (δ-CoV) [1] α- and β-CoVs are able to infect mammals, while γ-and δ-CoVs tend to infect birds [1]. HCoV-229E, HCoV-NL63 (α-CoVs) and HCoV-OC43, HCoV-HKU1 (β-CoVs) have crossed the species barriers from their bat reservoirs via various intermediate hosts to humans, and caused mild endemic infections of the upper respiratory tract such as common colds [2]. However, in recent years, several epidemic β-CoVs which were associated with severe acute respiratory syndrome (SARS) such as SARS-CoV-1, and middle east respiratory syndrome (MERS) such as MERS-CoV were considered as potential emergent pathogens for global pandemics [3, 4]. Most recently novel coronavirus (NCoV-19) also known as SARS-CoV-2 (β-CoV) which shows 96% genomic similarity with bat SARS-like coronavirus strain, BatCov RaTG13 have created the currently ongoing pandemic of coronavirus disease 2019. (COVID-19) [5, 6].

SARS-CoV-2 has a round or elliptic shape, often pleomorphic with a diameter of approximately 60–140 nm, and a nucleocapsid core surrounded by a lipid bilayer

IntechOpen

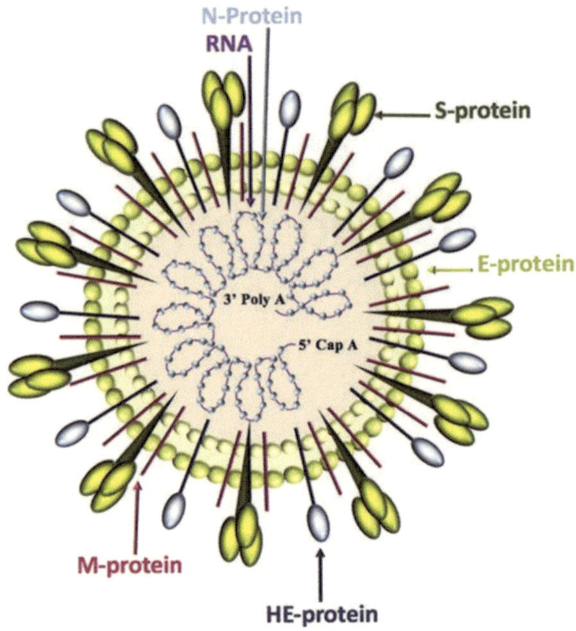

Figure 1.
Coronavirus structure. (Adapted from Fehr et al. [7]).

envelope (**Figure 1**) [7]. The nucleocapsid core contains the viral genome, single-stranded, non-segmented, positive-sense RNA which has a 5′ cap and a 3′ poly-A tail with a length of ~26.4 to ~31.7 kilobase (kb) complexed with the structural nucleocapsid (N) proteins (**Figure 1**) [7]. The lipid bilayer envelope which is taken by budding of RNA/nucleocapsid complex into the lumen of the ERGIC (endoplasmic reticulum (ER)–Golgi intermediate compartment) has the other structural glycoproteins including the spike (S) protein, the membrane (M) protein, the envelope (E) protein, and a fifth protein called hemagglutinin-esterase (HE) protein which binds to the terminal sialic acid residues on the host cell membrane glycoproteins and it manifests acetyl-esterase activity for the egress of SARS CoV-2 (**Figure 1**) [7, 8]. S protein, ~180 kDa glycoprotein is initially cleaved by the host serine protease furin resulting non-covalently linked transmembrane S2 subunit and a protruding extracellular S1 subunit during the intracellular maturation in the trans-Golgi-network [9]. Plasma membrane-exposed or secreted furin also cleaves S protein during entry of the virus resulting S1/S2 protomers which appear as mushroom-like trimers on the viral envelope (**Figure 1**) [9]. Each of the protomers can have an open or closed conformation. The "open" conformation of S1 exposes the receptor binding domain, (RBD) containing receptor binding motif (RBM) which shows increased binding affinity with angiotensin II-converting enzyme (ACE2) [10]. Currently circulating SARS-CoV-2 variant has a S1 D614G mutation with N-linked glycosylation sites N165 and N234 which favor the open conformation resulting the SARS-CoV-2 D614G variant more infectious [11]. SARS-CoV2 is also unique for having a proline residue between S1 and S2 subunits which leads to the formation of a turn/stem-loop structure resulting O-linked glycosylation at the cleavage site residues S686, S673, and T678 [12]. Following the binding of the amino- (N) terminal of S1 subunit RBM to ACE2, transmembrane protease serine 2 (TMPRSS2) and furin-mediated proteolytic activation/cleavage between the carboxyl- (C) terminal of S1 and N-terminal of S2 subunits results conformational change and fusion of the virus envelope and host cell membrane via C- terminal of S2 subunit, and this process delivers the virus genome into the host cell (**Figure 2**) [8]. SARS-CoV-2 also

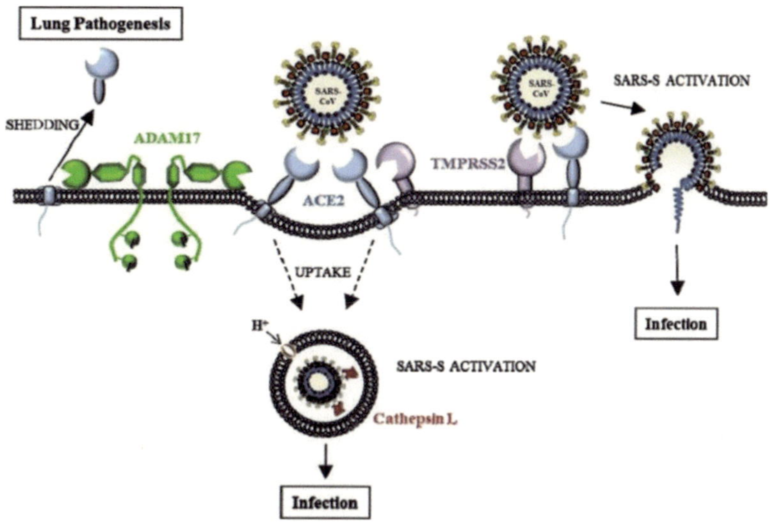

Figure 2.
Role of host cell proteases in the cellular entry of SARS-CoV. (Adapted from Heurich et al. [8]).

enters the host cell via receptor-mediated endocytosis (**Figure 2**) [8]. Upon binding of N terminal of S1 subunit RBM to ACE2 virion is taken up into the endosome, where S1 and S2 subunits are cleaved and activated by the pH-dependent cysteine protease, cathepsin L (**Figure 2**) [8]. Conformational change of the S1and S2 subunits allows fusion of the virus envelope with the endosomal membrane and release of the viral genome into the cytoplasm (**Figure 2**) [8]. ADAM17 (a disintegrin and metalloprotease 17) also known as tumor necrosis factor-α (TNF- α) converting enzyme, (TACE) is a membrane protease involved in the endogenous shedding of ACE2 complexed with S1 subunit RBD from the cell membranes (**Figure 2**) [8]. ADAM17-dependent ACE2 shedding is believed to promote lung pathogenesis [8, 13]. ACE2, is an essential component of the regulatory mechanism of the renin-angiotensin-aldosterone system (RAAS), and this pathway is considered to interact with the pathophysiology of COVID-19. In this chapter, we will discuss the key role of RAAS in the pathogenesis of Covid-19.

2. The renin: angiotensin: aldosterone system and pathophysiology of Covid 19

The RAAS is an important hormonal homeostatic mechanism of the body that involves the liver, kidneys, lungs and adrenal glands which plays a critical role in the regulation of blood pressure, fluid/electrolyte balance, systemic and pulmonary vascular resistance and vascular remodeling [14]. The function of the RAAS is mainly regulated by angiotensinogen, prorenin, renin, angiotensin I, (Ang I) angiotensin II, (Ang II) aldosterone, angiotensin 1–7, (Ang 1–7) angiotensin 1–9, (Ang 1–9) angiotensin I-converting enzyme, (ACE) and ACE2 (**Figure 3**) [15]. Prorenin, the precursor of renin, is proteolytically activated in the kidney by neuroendocrine convertase 1 (proprotein convertase 1) or cathepsin B, and nonproteolytically in many tissues by the renin/prorenin receptors [14]. Renin is produced by the juxtaglomerular cells in response to sympathetic nervous system (SNS) stimulation, hypotension, decreased cardiac output (CO) and renal perfusion pressure, decreased distal tubular sodium and chloride concentration and dehydration (**Figure 3**) [15]. Angiotensinogen, synthesized in the liver is an α-2-globulin, a member of the serpin family of proteins, but unlike the other serpins it is not known to inhibit proteases (**Figure 3**) [14, 15].

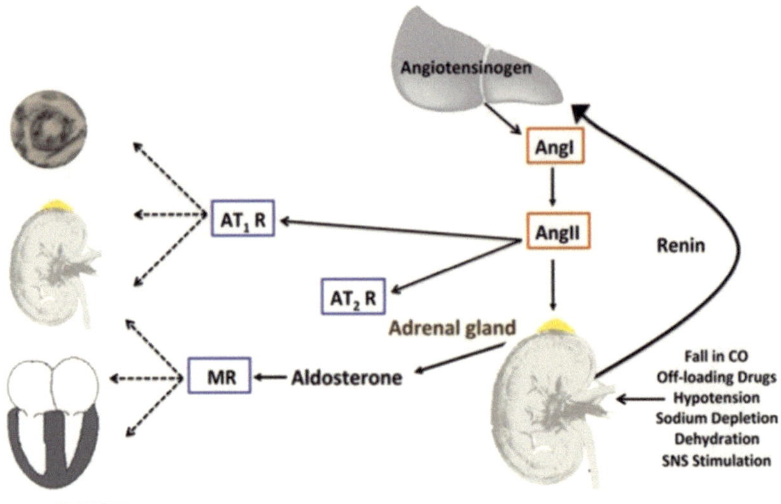

Figure 3.
The renin-angiotensin-aldosterone system (Adapted from Ames et al. [15]).

It has an elongated N-terminus as a substrate for renin which cleaves 10 N-terminus amino acids from angiotensinogen and creates the decapeptide Ang I (**Figure 3**) [15]. Ang I is considered to have no direct biological activity other than being a precursor to Ang II which is synthesized by ACE through removal of two C-terminal residues from Ang I (**Figure 3**) [15, 16]. ACE belongs to the M2 gluzincin family of metalloproteinases, zinc-dependent peptidyldipeptidase and it exists in two forms, somatic ACE and testicular ACE [16]. Both are derived from the same gene, controlled by alternative promoters [17]. Testicular ACE is considered to play a role in male fertility and sperm physiology [18]. Somatic ACE (ACE) is expressed in high amounts by the vascular endothelium of the lungs, renal proximal tubular epithelium and ciliated intestinal epithelium [17]. ACE mRNA expression has also been identified in different cell types and tissues including macrophages, dendritic cells, (DC) choroidal plexus and brain [19, 20]. The ACE gene promoter has been shown to harbor CpG islands which regulate ACE gene expression during inflammation via TNF-α, dependent hypermethylation resulting a decrease in cellular ACE activity [21, 22]. ACE is an integral membrane protein, which can be also cleaved by ACE secretases to produce a circulating form of the enzyme [23]. This soluble ACE activity is shown to be inhibited by an endogenous inhibitor which restricts ACE mediated Ang I conversion in the systemic circulation irrespective to the concentration of the circulating ACE that confines Ang II mediated responses in the tissues [24]. Changes in ACE expression have been shown to have minimum effect on blood pressure due to renin-mediated compensation of Ang I and its bioactive endogenous byproduct angiotensin 1–12 (Ang 1–12) which is more specialized for controlling blood pressure than Ang II [25, 26]. Ang II is considered to be mostly associated with innate and adaptive immunity, oxidative stress, inflammation and fibrosis [27, 28].

The Ang II receptors, (ATR1) and (ATR2), are a class of G protein-coupled receptors sharing a sequence identity of ~30%, but having a similar affinity for Ang II, which is their main ligand (**Figure 3**) [15, 16]. ATR2 stimulates the G protein-coupled receptor Gi subunit, and primarily inhibits the cAMP-dependent pathway by inhibiting adenylyl cyclase activity and decreasing the production of cAMP from ATP, which in turn results decreased activities of the cAMP-dependent protein kinases [29]. The downstream signaling pathways of these inhibitory processes lead to the modulation of protein kinase A (PKA), protein kinase C (PKC) and mitogen-activated protein kinase (MAPK) pathways, eventually inhibiting inflammation and growth-specific functional processes mostly affecting cardiac and vascular tissues [30]. ATR2 activation was reported to induce the activation of peroxisome proliferator-activated receptor, (PPARγ) a powerful anti-inflammatory factor in the post-ischemic cardiac tissue of rabbits that was accompanied by a down-regulation of MAPKs p42/44 [31]. ATR1 which stimulates the G protein-coupled receptor Gq protein alpha subunit on the vascular smooth muscle cell membranes which in turn activates an inositol triphosphate (IP3)-dependent mechanism leading to increase intracellular calcium levels and vasoconstriction (**Figure 3**) [15, 27]. Ang II stimulates aldosterone secretion from the adrenal gland cortex (**Figure 3**) [15]. Aldosterone increases sodium, chloride and bicarbonate reabsorption coupled with potassium and hydrogen excretion from the distal convoluted tubules, and amplifies the pathophysiologic effects of Ang II in the heart, kidney and vasculature via acting on the mineralocorticoid (MR) receptors (**Figure 3**) [15, 24]. More importantly, aldosterone was associated with inflammation via ER unfolded protein responses, mitochondrial dysfunction, as well as increased synthesis of pro-inflammatory cytokines such as interleukin 6. (IL-6) [32, 33]. It was also disclosed that activation of ATR1-receptors promotes Ang II-induced reactive oxygen species (ROS) generation, inflammation and angiogenesis via stimulating the Nicotinamide adenine dinucleotide phosphate (NADPH) oxidase, (NOX) nuclear factor kappa-light-chain-enhancer of activated B cells, (NF-κB) extracellular signal-regulated kinases, (ERK1/2) MAPK and signal transducer and activator of transcription 1 (STAT1) pathways [34, 35]. NOX is the best known non-mitochondrial source of ROS generation [36]. p22phox subunit of NOX is required for activating, stabilizing and/or regulating NOX homologs [36]. Ang II reportedly induces oxidative stress by elevating the expression of p22phox [37]. ROS are considered to oxidize membrane phospholipids, proteins and nucleic acids, and lead to tissue hypertrophy and inflammation mainly in the alveolar epithelial cells, endothelium and heart via triggering the synthesis of adhesion molecules including intercellular adhesion molecule 1, (ICAM-1) vascular cell adhesion protein 1, (VCAM-1) monocyte chemoattractant protein 1, (MCP-1) and macrophage colony stimulating factor. (M-CSF) [34, 35]. Moreover, Ang II stimulates production of ROS from NO leading to depletion of NO causing further injury to blood vessels [14]. Additionally, C- reactive protein (CRP) induces ATR1 transcription and translation as well as enhanced ATR1 levels in blood vessel wall [14, 35]. Heat shock proteins (HSPs) have been found to be a regulator of NF-κB cascade in inflammation induced by Ang II via activation of the inhibitor of nuclear factor kappa B (IκB) kinase (IKK) complex and phosphorylation of IκBα. This process leads to ubiquitination and degradation of IκBα, and permits NF-κB translocation to the nucleus. NF-κB stimulates the transcription of proinflammatory cytokines including TNF-α, IL-6, IL-8, MCP-1, and cyclooxygenase [38]. Cyclooxygenase 1-derived prostaglandin E2 and prostaglandin E2 type 1 receptors are considered to play a role in Ang II-dependent hypertension via AT1R/phospholipase A2 pathway which promotes ROS production coupled with $Ca2+$ influx [39]. TNF-α, primary substrate for ADAM17 is cleaved and released from the cell membrane forming the soluble TNF-α which in turn binds and activates TNF-α receptors on the cell surfaces [40]. ADAM17 activity is upregulated by the binding of soluble TNF-α to its receptors, and also via the ATR1/Ang II axis [40]. Ang II enhances activation of MAPK cascades including ERK1/2, c-Jun N-terminal kinase (JNK) and ERK5 via ATR1 resulting increased synthesis of matrix metalloproteinase-2 (MMP-2) which amplifies the inflammation associated with the proinflammatory cytokines and results to angiogenesis, widespread disruption of endothelial barriers and cardiac abnormalities [41]. Activation of STAT1/STAT2 downstream pathway via Ang II – ATR1 binding stimulates interferon-stimulated genes (ISG) expression by inducing the interferon-stimulated response element

(ISRE) promoter and increases the maturation and activation of the antigen presenting cells, natural killer cells and T-box expressed in T cells (T-bet) cells which are bridging between innate and adaptive immunity and leading to autoimmune reactions and cardiovascular diseases [42]. Ang II-induced vasoconstriction and inflammatory endothelial cell injury have been associated with accelerated thrombus development in the arteries, veins, and capillaries via activation of different components of the coagulation cascade [42, 43]. ACE2 is a zinc-carboxypeptidase consisting of 805 amino acids with an extracellular N-terminal domain, transmembrane (TM) domain and an intracellular C-terminal tail (**Figure 4**) [44]. The zinc-binding motif (HEMGH) is located within the carboxypeptidase domain which also recognizes RBM of S1 subunit of SARS-CoV-2 (**Figure 4**) [44]. Collectrin domain is the site for ACE2 shedding with ADAM17 and TMPPSS2, and it is crucial for interacting with neutral amino acid transporters (**Figure 4**) [44]. ACE2, which has 42% identical nucleotide sequence with ACE indicating that the two genes, ACE2 and ACE arise through duplication [45]. ACE2 is expressed in a diverse group of cells including the oral, nasal, type II lung alveolar, tongue and esophageal epithelial cells, enterocytes, endothelial cells, cardiomyocytes, arterial smooth muscle cells in most organs, cortical neurons and glia, renal tubules, ductal cells, bladder urothelial cells and male reproductive cells [44–46]. ACE2 cleaves the carboxyl (C)-terminal amino acid phenylalanine from Ang II, (Asp-Arg-Val-Tyr-Ile-His-Pro-Phe) and hydrolyses it into the vasodilator Ang (1–7), (Asp-Arg-Val-Tyr-Ile-His-Pro) a ligand for the G-protein coupled receptor Mas receptor (MasR) (**Figure 5**) [47]. ACE2 also converts Ang I to Ang 1–9 that can be further hydrolyzed to Ang 1–7 by the action of neprilysin, (Nep) which is a zinc-dependent metalloprotease, and cleaves peptides at the N-side of hydrophobic residues (**Figure 5**) [47, 48]. Nep also directly converts Ang I to Ang 1–7 which is further enzymatically decarboxylated to alamandine, a ligand for Mas-related G-protein-coupled receptor, member D (MrgD) (**Figure 5**) [47, 49]. Activation of Mas and MrgD receptors upon binding with Ang 1–7 and alamandine respectively promote anti-inflammatory responses via increasing the levels of anti-inflammatory cytokines IL-4 and IL-10 which have been disclosed in macrophages and microglial cells [50, 51]. ACE2 is considered as a key modulator of the RAAS via regulating physiological and pathological functions of cardiovascular, renal and pulmonary systems via counterbalancing the hypertensive, vasoconstrictor, hypertrophic and inflammatory effects of ACE [47, 52]. Thus, the ratio of Ang II/ACE2 plays an important role in the pathogenesis of several diseases including Covid-19 [47, 52, 53]. Two forms of ACE2 have been reported [44].

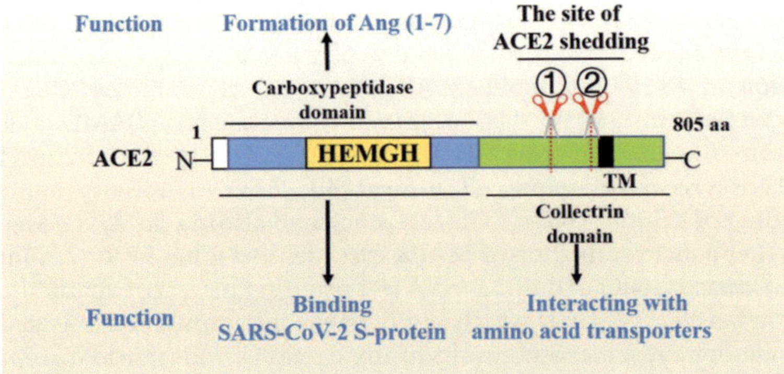

Figure 4.
Domain structure and function of angiotensin-converting enzyme 2. (Adapted from Bian et al. [44]).

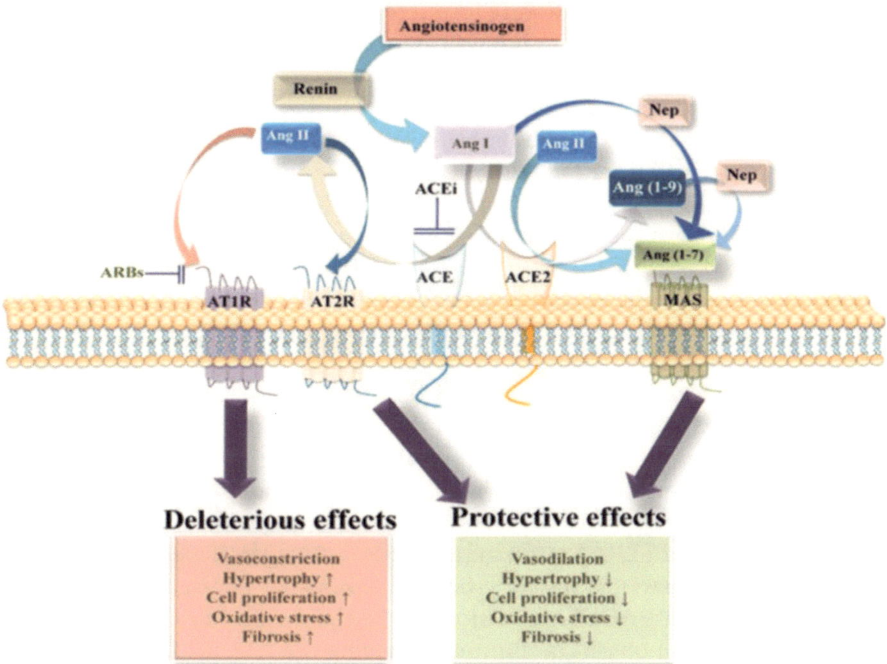

Figure 5.
A classic model of RAAS showing deleterious and protective effects. (Adapted from Gul et al. [47]).

The full-length membrane-bound ACE2 (mACE2) is located on the apical surface of epithelial cells, differently from ACE, which is located between the apical and basolateral membranes in polarized cells [19, 44]. S1 trimeric subunit RBM of SARS-CoV-2 binds to the widely expressed mACE2 extracellular domain (**Figures 2** and **4**) [8, 44]. The second form soluble ACE2 (sACE2) is shed into the circulation via Ang II/AT1R/ADAM17 axis (**Figure 2**) [8, 54, 55]. Although the levels of sACE2 may be increased in plasma or urine in some pathological processes, such as hypertension, the expression levels of mACE2 is not affected [45]. The majority of ACE2 is membrane-bound, and it has a compensatory balancing effect on the RAAS [49]. Ang II counterbalances the number of mACE2 via cellular internalization through endocytosis and degradation in the lysosomes, thus inhibits the antioxidative, anti-inflammatory, anti-hypertrophic and antifibrotic effects of ACE2 [55]. ACE2 inactivity via shedding, cellular internalization and degradation at early stages of COVID-19 might have a disturbing effect on the RAAS homeostasis which leads to increased vascular permeability, fluid accumulation in the extra-alveolar spaces, oxidant/antioxidant imbalance and impaired tissue repair [56, 57]. COVID-19 patients were found to suffer more frequently from severe endothelial injury due to the membrane damage by binding of SARS-CoV-2 [58]. Widespread thrombosis with microangiopathies were reported, and the COVID-19 patients were found 9 times more likely to experience alveolar capillary microthrombi, and 2.7 times more likely to experience intussusceptive angiogenesis than the patients with flu [59]. This phenomenon of thrombosis and other vascular events in COVID-19 were considered more likely associated with abundance of ACE2 on the endothelial cell membranes which permit SARS-CoV-2 infection along the endothelium resulting endothelial damage, complement activation, release of Von-Willebrand factor from the endothelial cells, hypercoagulability and microthrombi formation [58, 59]. ACE2 expression is regulated by genetic and epigenetic factors, body mass index, inflammatory cytokines, cigarette smoking, sex hormones and aging. ACE2 expression in different tissues across human individuals were found to be high in Asian ethnic groups [60]. The upregulation of ACE2 expression was associated with the decline in the levels of estrogen and androgen, aging, inflammation and cigarette smoking [60, 61].

Cis-elements in the proximal promoters of ACE2 genes have binding sites for canonical interferon- (IFN) dependent transcription factors including ISRE/STAT1, interferon regulatory factor 1 (IRF1), IRF3/7 and IRF8 [62]. Type I IFNs, and to a lesser extent type II and type III IFNs have been shown to upregulate ACE2 expression especially in the human upper airway basal cells and bronchial cells [63]. Higher enrichment of ISRE/STAT1/3 and/or IRF3/7 binding sites were detected in single cell RNA-sequence data sets from the nasal epithelium and upper airway goblet secretory cells of the patients with the severe manifestations of COVID-19 suggesting dual roles of ACE2 in the pathogenesis of SARS- COV-2 [64]. First, ACE2 serves as an innate immune receptor for SARS-CoV-2 which might compete with Ang II for the binding sites located at the carboxypeptidase domain of ACE2, and SARS-COV-2 gains access into the cells via ACE2 receptor. The similar functional innate protein–protein interactions between the human toll-like receptors (TLRs) TLR1, TLR4, and TLR6 with a binding energy values of −57.3, −120.2, −68.4 respectively, being the TLR4-S protein interaction strongest have been demonstrated by the molecular docking [65]. Secondly, innate IFN responses against SARS-CoV-2 upregulate ACE2 expression on the cell membranes which augments anti-inflammatory responses via counter-balancing the effects of Ang II, but also allows further cellular entry of SARS-CoV-2. However; the balancing arm of the RAAS functioning as ADAM17 mediated ACE2 shedding and ADAM17 mediated TNF-α activation/hypermethylation of the CpG islands at the ACE gene promoter eventually decrease cellular entry of SARS-CoV-2 and ACE expression.

In summary, there are couple of balancing and counter-balancing factors existing in the pathogenesis of Covid-19. The main counter-balancing arm is between Renin/ACE/Ang II/ATR1 and ATR2/ACE2/Mas/MrgD. The second counter-balancing arm is between the innate immune responses including SARS- CoV-2 induced IFN response/increased membrane expression of ACE2, and ADAM17 mediated ACE2 shedding activated by ATR1/Ang II and TNF-α mediated downregulation of ACE expression. The second arm has its own feedback via TNF-α. However, the main arm counter-balancing the protective and deleterious effects of RAAS which is being abused by SARS- CoV-2 via competing with Ang II for binding to the ACE2 receptor plays a crucial role in the pathogenesis of Covid-19 via the unopposed effects of Ang II including oxidative stress, inflammation, stimulation of innate and adaptive immunity via T-bet cells, thrombosis, angiogenesis and fibrosis.

Author details

Ozlem G. Sahin
Department of Pathology Microbiology and Immunology, New York Medical College, Valhalla, NY, USA

*Address all correspondence to: osahin@student.nymc.edu

References

[1] Gorbalenya A.E., Snijder E.J., Spaan W.J.M. Severe acute respiratory syndrome coronavirus phylogeny: toward consensus. J. Virol. 2004;78: 7863-7866.

[2] Liu D.X., Liang J.Q., Fung T.S. Human Coronavirus-229E, -OC43, -NL63, and -HKU1. Reference Module in Life Sciences. 2020 : B978-0-12-809633-8.21501-X. Published online 2020 May 7.

[3] Summary of probable SARS cases with onset of illness from 1 November 2002 to 31 July 2003. https://www.who.int/csr/sars/country/table2004_04_21/en/ (Based on data as of the 31 December).

[4] WHO MERS-CoV Global Summary and risk assessment. 2006. https://www.who.int/emergencies/mers-cov/mers-summary-2016.pdf (Based on data as 6 December).

[5] Zhou P., Yang X. L., Wang X. G., et al. A pneumonia outbreak associated with a new coronavirus of probable bat origin. Nature 2020;579,270-273.

[6] WHO Coronavirus Disease (COVID-19). Dashboard. Accessed 15 July 2020.

[7] Fehr A.R., Perlman S. Coronaviruses: an overview of their replication and pathogenesis. Methods Mol. Biol.2015; 1282:1-23.

[8] Heurich A., Hofmann-Winkler H., Gierer S., et al. TMPRSS2 and ADAM17 cleave ACE2 differentially and only proteolysis by TMPRSS2 augments entry driven by the severe acute respiratory syndrome coronavirus spike protein. J. Virol.2014;88, 1293-1307.

[9] Hoffmann M, Kleine-Weber H, Pohlmann S. A Multibasic Cleavage Site in the Spike Protein of SARS-CoV-2 Is Essential for Infection of Human Lung Cells. Mol Cell. 2020;78:779-784.

[10] Mercurio I., Tragni V., Busto F., et al. Protein structure analysis of the interactions between SARS-CoV-2 spike protein and the human ACE2 receptor: from conformational changes to novel neutralizing antibodies. Cell Mol Life Sci. 2020; 4 :1-22.

[11] Gobeil S.M.C., Janowska K., McDowell S., et al. D614G Mutation Alters SARS-CoV-2 Spike Conformation and Enhances Protease Cleavage at the S1/S2 Junction. Cell Rep. 2020 Dec 26:108630. doi: 10.1016/j.celrep.2020.108630 [Epub ahead of print]

[12] Zhao P., Praissman J.L, Grant O. C., et al. Virus-Receptor Interactions of Glycosylated SARS-CoV-2 Spike and Human ACE2 Receptor. Cell Host Microbe. 2020;28:586-560.

[13] Salamanna F., Maglio M., Landini M. P., et al. Body Localization of ACE-2: On the Trail of the Keyhole of SARS-CoV-2. Front Med (Lausanne) 2020; 7:594495.

[14] Forrester S. J., Booz G. W., Sigmund C. D., et al. Angiotensin II Signal Transduction: An Update on Mechanisms of Physiology and Pathophysiology. Physiol Rev. 2018;98: 1627-1738.

[15] Ames M. K., Atkins C.E., Pitt B. The renin-angiotensin-aldosterone system and its suppression. J Vet Intern Med. 2019;33:363-382.

[16] Sparks M. A., Crowley S. D., Gurley S. B., et al. Classical Renin-Angiotensin System in Kidney Physiology. Compr Physiol. 2014;4:1201-1228.

[17] Harrison C., Acharya R. ACE for all – a molecular perspective. J Cell Commun Signal. 2014;8:195-210.

[18] Gianzo M., Subiran N. Regulation of Male Fertility by the Renin-Angiotensin System. Int J Mol Sci. 2020;21:7943.

[19] Bruneval P., Hinglais N., Alhenc-Gelas F., et al. Angiotensin I converting enzyme in human intestine and kidney. Ultrastructural immunohistochemical localization. Histochemistry. 1986;85:73-80.

[20] Skidgel R.A., Erdös EG. Biochemistry of angiotensin converting enzyme, in the Renin-Angiotensin System. Vol. 1 (Robertson JIS, Nicholls MG. eds) pp 10.1-10.10, Gower Medical Publishers, London, 1993.

[21] Rivière G., Lienhard D., Andrieu T., et al. Epigenetic regulation of somatic angiotensin-converting enzyme by DNA methylation and histone acetylation. Epigenetics. 2011;6:478-489.

[22] Mudersbach T., Siuda D., Kohlstedt K., et al. Epigenetic control of the angiotensin-converting enzyme in endothelial cells during inflammation. PLoS One. 2019;14: e0216218.

[23] Oppong S. Y., Hooper N. M. Characterization of a secretase activity which releases angiotensin-converting enzyme from the membrane. Biochem J. 1993;292:597-603.

[24] Fagyas M., Uri K., Siket I. M., et al. New Perspectives in the Renin-Angiotensin-Aldosterone System (RAAS) I: Endogenous Angiotensin Converting Enzyme (ACE) Inhibition. PLoS One. 2014;9:e87843.

[25] Smithies O., Kim H.S., Takahashi N., et al. Importance of quantitative genetic variations in the etiology of hypertension. Kidney Int. 2000;58:2265-2280.

[26] Musini V. M., Lawrence K. A. K., Fortin P. M., et al. Blood pressure lowering efficacy of renin inhibitors for primary hypertension. Cochrane Database Syst Rev. 2017;2017:CD007066.

[27] Nehme A., Zouein F. A., Zayeri Z. D., et al. An Update on the Tissue Renin Angiotensin System and Its Role in Physiology and Pathology. J Cardiovasc Dev Dis. 2019;6:14.

[28] Kryukova O. V., Tikhomirova V. E., Golukhova E. Z., et al. Tissue Specificity of Human Angiotensin I-Converting Enzyme. PLoS One. 2015;10:e0143455.

[29] Tadevosyan A., Letourneau M., Folch B., et.al. Photoreleasable ligands to study intracrine angiotensin II signalling. J Physiol. 2015;593: 521-539.

[30] Namsolleck P., Recarti C., Foulquier S., et al. AT2 receptor and tissue injury: therapeutic implications. Curr Hypertens Rep. 2014;16:416.

[31] Molavi B., Chen J., Mehta J.L. Cardioprotective effects of rosiglitazone are associated with selective overexpression of type 2 angiotensin receptors and inhibition of p42/44 MAPK. Am. J. Physiol. Heart Circ. Physiol. 2006;291:H687–H693.

[32] Bordt E. A., Smith C. S., Demarest T.G., et al. Mitochondria, Oxytocin, and Vasopressin: Unfolding the inflammatory protein response. Neurotox Res. 2019; 36:239-256.

[33] Luther J.M., Gainer J.V., Murphey L.J., et al. Angiotensin II induces interleukin-6 in humans through a mineralocorticoid receptor-dependent mechanism. Hypertension. 2006;48:1050-1057.

[34] Montecucco F., Pende A., Mach F. The Renin-Angiotensin System Modulates Inflammatory Processes in Atherosclerosis: Evidence from Basic Research and Clinical Studies. Mediators Inflamm. 2009;2009:752406.

[35] Husain K., Hernandez W., Ansari R. A., et al. Inflammation, oxidative stress and renin angiotensin system in atherosclerosis. World J Biol Chem. 2015;6:209-217.

[36] Bedard K., Krause K. H. The NOX family of ROS-generating NADPH oxidases: physiology and pathophysio logy. Physiol Rev. 2007;87:245-313.

[37] Qui Y., Tao L., Lei C., et al. Downregulating p22phox ameliorates inflammatory response in Angiotensin II-induced oxidative stress by regulating MAPK and NF-κB pathways in ARPE-19 cells. Sci Rep. 2015;5:14362.

[38] Lee K. H., Jang Y., Chung J.H. Heat shock protein 90 regulates IκB kinase complex and NF-κB activation in angiotensin II-induced cardiac cell hypertrophy. Exp Mol Med. 2010;42: 703-711.

[39] Wang G., Sarkar P., Peterson J.R., et al. COX-1-derived PGE2 and PGE2 type 1 receptors are vital for angiotensin II-induced formation of reactive oxygen species and Ca2+ influx in the subfornical organ. Am J Physiol Heart Circ Physiol. 2013;305: H1451–HH146.

[40] Gheblawi M., Wang K., Viveiros A., et al. Angiotensin-Converting Enzyme 2: SARS-CoV-2 Receptor and Regulator of the Renin-Angiotensin System: Celebrating the 20th Anniversary of the Discovery of ACE2.Circ Res. 2020;126:10.1161/CIRCRESAHA.120. 317015.

[41] Wang C., Qian X., Sun X., et al. Angiotensin II increases matrix metalloproteinase 2 expression in human aortic smooth muscle cells via AT1R and ERK1/2. Exp Biol Med. (Maywood) 2015;240:1564-1571.

[42] Kossmann S., Schwenk M., Hausding M., et al. Angiotensin II-induced vascular dysfunction depends on interferon-γ-driven immune cell recruitment and mutual activation of monocytes and NK-cells. Arterioscler Thromb Vasc Biol. 2013;33: 1313-1319.

[43] Daiber A., Steven S., Weber A., et al. Targeting vascular (endothelial) dysfunction. Br J Pharmacol. 2017;174:1591-1619.

[44] Bian J., Li Z. Angiotensin-converting enzyme 2 (ACE2): SARS-CoV-2 receptor and RAS modulator. Acta Pharm Sin B. 2020;doi: 10.1016/j.apsb.2020.10.006.

[45] Tipnis S.R., Hooper N.M., Hyde R., et al. A human homolog of angiotensin-converting enzyme. Cloning and functional expression as a captopril-insensitive carboxypeptidase. J Biol Chem. 2000;275:33238-33243.

[46] Chen J., Jiang Q., Xia X., et al. Individual variation of the SARS-CoV-2 receptor ACE2 gene expression and regulation. Aging Cell. 2020;19 e13168.

[47] Gul R., Kim U.H., Alfadda A.A. Renin-angiotensin system at the interface of COVID-19 infection. Eur J Pharmacol. 2021;890:173656.

[48] Donoghue M., Hsieh F., Baronas E., et.al. A novel angiotensin-converting enzyme-related carboxypeptidase (ACE2) converts angiotensin I to angiotensin 1-9. Comparative Study Circ Res. 2000;87:E1-E9.

[49] Santos R.A.S., Ferreira A.J., Verano-Braga T., et al. Angiotensin-converting enzyme 2, angiotensin-(1-7) and Mas: New players of the renin–angiotensin system. J. Endocrinol. 2012;216:R1–R17.

[50] Santuchi M. C., Dutra M. F., Vago J.P., et al. Angiotensin-(1-7) and Alamandine Promote Anti-inflammatory Response in Macrophages In Vitro and In Vivo. Mediators Inflamm. 2019;2019:2401081.

[51] Liu M. Shi P., Sumners C. Direct anti-inflammatory effects of angiotensin-(1-7) on microglia. J Neurochem. 2016;136:163-171.

[52] Zamai L. The Yin and Yang of ACE/ACE2 Pathways: The Rationale for the

Use of Renin-Angiotensin System Inhibitors in COVID-19 Patients. Cells. 2020;9:1704.

[53] Ocaranza M.P., Riquelme J.A., García L., et al. Counter-regulatory renin–angiotensin system in cardiovascular disease. Nat. Rev. Cardiol. 2019;17:116-129.

[54] Jia H.P., Look D.C., Tan P., Shi L., et al. Ectodomain shedding of angiotensin converting enzyme 2 in human airway epithelia. Am. J. Physiol. Lung Cell. Mol. Physiol. 2009;297:L84–L96.

[55] Deshotels M. R., Xia H., Sriramula S., et al. Angiotensin-II mediates ACE2 Internalization and Degradation through an Angiotensin-II type I receptor-dependent mechanism. Hypertension. 2014;64:1368-1375.

[56] Hrenak J., Simko F. Renin–Angiotensin System: An Important Player in the Pathogenesis of Acute Respiratory Distress Syndrome. Int J Mol Sci. 2020;21:8038.

[57] Zipeto D., Palmeira J. F., Arganaraz G. A., et al. ACE2/ADAM17/TMPRSS2 Interplay May Be the Main Risk Factor for COVID-19. Front Immunol. 2020;11:576745.

[58] Ackermann M., Verleden S. E., Kuehnel M., et al. Pulmonary Vascular Endothelialitis, Thrombosis, and Angiogenesis in Covid-19. N Engl J Med. 2020;383:120-128.

[59] Ackermann M., Mentzer S., Kolb M., et al. Inflammation and intussusceptive angiogenesis in COVID-19: everything in and out of flow. Eur Respir J. 2020;56:2003147.

[60] Chen J., Jiang Q., Xia X., et al. Individual variation of the SARS-CoV-2 receptor ACE2 gene expression and regulation. Aging Cell. 2020;19:e13168.

[61] Smith J. C., Sausville E. L., Girsh S. V., et al. Cigarette Smoke Exposure and Inflammatory Signaling Increase the Expression of the SARS-CoV-2 Receptor ACE2 in the Respiratory Tract. Dev Cell. 2020;53:514-529.

[62] Sang E. R., Tian Y., Gong Y., et al. Integrate structural analysis, isoform diversity, and interferon-inductive propensity of ACE2 to predict SARS-CoV2 susceptibility in vertebrates. Heliyon. 2020;6:e04818.

[63] Busnadiego I., Fernbach S., Pohl M. O., et al. Antiviral Activity of Type I, II, and III Interferons Counterbalances ACE2 Inducibility and Restricts SARS-CoV-2. mBio. 2020;11:e01928-e01920.

[64] Ziegler C. G. K., Allon S. J., Nyquist S. K., et al. SARS-CoV-2 Receptor ACE2 Is an Interferon-Stimulated Gene in Human Airway Epithelial Cells and Is Detected in Specific Cell Subsets across Tissues. Cell. 2020;181:1016-1035.

[65] Choudhury A., Mukherjee S. In silico studies on the comparative characterization of the interactions of SARS-CoV-2 spike glycoprotein with ACE-2 receptor homologs and human TLRs. J Med Virol. 2020;17:10.1002/jmv.25987.